# Mechanobiology

*How Cells Read Force and Build Bodies*

Team Trex

Published by NobleTrex Press



For permissions and other inquiries, write to:

P.O. Box 3132, Framingham, MA 01701, USA

# Contents

# Introduction

If you were to take the complete human genome-the entire three-billion-letter code that defines a person-and dissolve it in a test tube of water, you would have a chemically perfect recipe for a human being. It would contain every instruction needed to build eyes, heart valves, neurons, and skin. Yet, if you watched that test tube for a thousand years, nothing would happen. The DNA would float there, an inert library of potential, waiting for a librarian that never arrives.

This is the great paradox of modern biology. For the last seventy years, we have been obsessed with the code. We have mapped the genome, cataloged the mutations, and convinced ourselves that life is fundamentally a chemistry set. We tend to view the cell as a tiny bag of reactions where protein A bumps into protein B to create effect C. But this view misses half the story.

A blueprint, no matter how detailed, cannot build a house. It cannot hammer a nail, pour concrete, or lift a beam against the pull of gravity. To turn a plan into a structure, you need physical force. You need pushing, pulling, tension, and compression.

Life is no different. Your cells are not merely chemical computers; they are engines. They are architects and construction workers. They crawl, they squeeze, they tear, and they stick. They feel the stiffness of the tissue around them and change their behavior based on whether they are sitting on soft brain matter or rigid bone. They have a sense of touch so refined it makes our fingertips seem clumsy by comparison.

This is the story of **mechanobiology**, the science of how physical forces shape life. It is an exploration of the invisible hand that sculpts us from a single egg into a complex organism, keeps our hearts beating, and, when it goes wrong, drives the progression of deadly diseases.

To understand why this matters, you don't need a microscope; you only need to move. Stand up. When you do, the bones in your legs undergo microscopic compression. This isn't just passive structural support; inside the bone, cells called osteocytes feel that squeeze. They detect the flow of fluid rushing through tiny canals in the mineral matrix. That physical shear stress tells them, "We are under load. Build more bone." This is why astronauts lose bone density in space. Without the relentless drag of gravity, their cells stop receiving the mechanical orders to maintain the skeleton. The chemistry is fine, but the physics is missing.

Consider the beat of your heart. For a long time, we thought the flow of blood was simply a delivery system for oxygen. We now know that the friction of blood sliding against the inner walls of your arteries is a language. The endothelial cells lining your vessels align themselves with this flow, like grass flattening in the wind. If that flow becomes turbulent-perhaps at a sharp bend in the artery-the mechanical signal becomes garbled. The cells become confused, inflamed, and eventually allow the buildup of plaques that lead to heart attacks. The disease begins not just with cholesterol, but with a disturbance in the flow.

Mechanobiology challenges us to look at the human body not just as a chemical factory, but as a mechanical structure that is constantly under stress. It asks a simple but profound question: How does a cell know where it is?

For decades, if you asked a biologist how a cell determines its fate, they would point to chemical signals-morphogens and growth factors. But in the early 20th century, the Scottish mathematician and biologist D'Arcy Wentworth Thompson argued for a different view. In his 1917 masterpiece, *On Growth and Form*, he suggested that biological forms are diagrammatic representations of the forces acting upon them. He saw the curves of a horn or the lattice of a sponge not as genetic predestination, but as the result of physical laws.

Thompson was ahead of his time-perhaps too far ahead. With the discovery of the DNA double helix in the 1950s, biology pivoted hard toward genetics. The physical nature of the cell was relegated to the background, viewed as mere housekeeping. The cytoskeleton-the cell's internal scaffolding-was seen as a static frame, not the dynamic, force-generating machine we now know it to be.

The renaissance of physical biology began only recently, driven by new tools that allowed us to poke, prod, and listen to cells. We can now measure the traction force a single cell exerts on its substrate-a microscopic grip strength test. We have discovered that cells are not fluid-filled balloons, but tensegrity structures, balancing internal tension against external compression.

One of the most striking revelations came in 2006, when Adam Engler, Dennis Discher, and their colleagues demonstrated something that shocked the field. They took identical stem cells and placed them on gels of varying stiffness. They added no special chemical cocktails. They simply changed the "feel" of the floor. Cells placed on a soft, brain-like gel became neurons. Those on a medium-stiffness, muscle-like gel became muscle precursors. And those on a rock-hard, bone-like surface turned into osteoblasts. The cells were reading the physics of their environment and deciding their destiny based on touch alone.

This discovery overturned the idea that the genome is the sole dictator of identity. It turns out the genome is a piano, but mechanical force is the pianist.

As we venture into this hidden world, we must first learn the language of the cell. It is a language of stress and strain, of viscosity and elasticity. We will see how life exists in a perpetual state of "viscoelasticity"-behaving like a solid when you hit it fast, but flowing like a liquid if you push it slowly. This property is what allows your tissues to absorb the shock of a jump yet slowly reshape themselves during growth.

Deep inside the cell, we will find the machinery responsible for this magic. We will meet the motor proteins-kinesins and myosins-that walk along tightropes of actin and tubulin, hauling cargo and generating the tension that keeps the cell alive. This is the "inner scaffold," a structure that is constantly building and destroying itself, consuming energy to maintain a state of readiness. It is here that the cell generates the forces necessary to divide, to crawl toward a wound, or to squeeze through the tight gaps in a capillary.

But a machine that generates force is useless if it cannot feel the forces acting upon it. This is where the concept of *mechanotransduction* comes in-the translation of mechanical force into chemical signals. It is the Rosetta Stone of mechanobiology.

We will explore the exquisite sensitivity of molecular sensors like the Piezo ion channels, identified by Ardem Patapoutian's lab, which earned a Nobel Prize in 2021. These proteins sit in the cell membrane like trapdoors. When the membrane stretches, even by a nanometer, the doors snap open, letting charged ions rush in and alerting the cell to the pressure. It is because of these sensors that you can feel the brush of a feather on your arm or the fullness of your bladder.

Yet, this sensitivity is not just for sensation; it is for construction. We will examine how cells hold hands, linking their internal skeletons together via cadherins to form sheets of tissue. We will watch as they weave a complex tapestry around themselves called the extracellular matrix-a mesh of collagen and elastin that provides the structural integrity for our organs. This matrix is not dead scaffolding; it is an information storage system. By pulling on this matrix, cells communicate with their neighbors over long distances, tugging on the web like spiders to signal their presence.

The implications of this physical dialogue are staggering, particularly when we look at how bodies are built. An embryo is a masterpiece of origami. It starts as a flat sheet of cells that must fold, twist, and tube itself into a body. These movements are driven by coordinated waves of mechanical force. If you disrupt the tension, the genetic code remains perfect, but the embryo fails to form. The heart cannot loop; the neural tube cannot close. The genes provide the parts list, but mechanics provides the assembly instructions.

There is, however, a darker side to this story. If force creates form, then distorted forces can create disease. We will investigate the "rogue mechanics" of cancer. For a long time, oncologists have known that tumors feel harder than healthy tissue-it's why we perform self-exams for lumps. But we now know that this stiffness is not just a symptom; it is a driver. The rigid scar tissue around a tumor tricks the cancer cells into thinking they are in a wound that needs healing, prompting them to proliferate aggressively. Worse, we will see how cancer cells use their mechanical engines to drill through tissue layers, escaping into the bloodstream to metastasize. They are experts at squeezing through gaps that should be impassable, hijacking the body's own mechanical pathways.

This leads us to the problem of fibrosis-the scarring that destroys organs like the liver, lungs, and heart. Fibrosis is a mechanical vicious cycle: a stiff environment activates cells to produce more stiff matrix, which in turn activates more cells. Breaking this loop requires more than just drugs; it requires "mechanomedicine."

The future of medicine may well lie in physics. We are entering an era where we can engineer materials that speak to cells in their own language. We are designing implants with specific surface topographies that encourage bone growth without drugs. We are creating "organoids"-miniature organs in a dish-that pulse and function because we have learned how to recreate the mechanical environment of the womb. We are even exploring how to use sound waves and physical stretching to rehabilitate injured muscle or calm inflammation.

Writing about mechanobiology requires a shift in perspective. We are used to thinking of ourselves as biological entities, defined by our DNA. But we are equally physical entities, defined by gravity, friction, and tension.

As we journey through the chapters ahead, try to visualize the microscopic storms raging inside you. Right now, as your eyes scan this page, tiny molecular motors are ratcheting along protein cables to keep your retinal cells in shape. Your skin cells are pulling on one another to maintain the barrier that protects you from the air. Your arteries are pulsing, stretching and relaxing with a rhythm that has continued since before you were born.

We are made of soft matter, yet we stand upright. We are fluid, yet we hold our shape. We are written in code, but we are built by force.

Let's look under the hood.

Chapter 1

# The Invisible Sculptor: Forces That Shape Life

*Life is often described as a chemical script written in DNA, yet cells inhabit a volatile physical world governed by gravity, tension, and stiffness. This chapter reveals how physical forces-the push of fluids and the pull of fibers-act as invisible hands that sculpt embryos, maintain organs, and drive the very definition of being alive.*

## 1.1 The Second Genetic Code

A strand of DNA is often treated like destiny: a fixed script that, once read, produces a cell, a tissue, a body. The metaphor is powerful because it is partly true. Genes encode proteins, and proteins build and run cells. Yet even a flawless script cannot stage a play on an empty lot. Cells, like actors, need cues: where to stand, when to speak, how loudly to respond. Many of those cues are chemical—hormones, growth factors, nutrients. But some are physical. A cell lives inside a material world, pushed and pulled by its neighbors, pressed by fluid pressure, tugging on the fibers around it. And it does not merely endure those forces; it interprets them. The act of converting force into biochemical change is called *mechanotransduction*. If genes are the parts list, mechanotransduction is part of the instruction manual:

it tells the same genome which page to open.

The simplest way to feel the difference between genetic information and mechanical information is to imagine two identical seeds. Plant one in loose, dark soil; wedge the other into a crack in dry pavement. Their genomes are the same, but the physical context is not, and the plants will not be the same. Animals are no different, except the physics is more intimate: the "soil" is the tissue itself, and the organism supplies much of the force. Cells build the scaffolding that surrounds them, then read it, then rebuild it again—a continual loop in which material becomes message.

The idea that the environment can reach into the nucleus and change gene expression is not new. What is newer is the realization that the environment speaks in a mechanical dialect as well as a chemical one. A foundational clue came from work on epithelial tissues and the extracellular matrix: the meshwork of proteins and sugars outside the cell that provides both structure and signals. Mina Bissell and colleagues, in the early 1980s, argued that the extracellular matrix is not just glue or packaging but an active regulator of gene expression. Their phrase *dynamic reciprocity* captured a feedback loop that now sits near the heart of mechanobiology: cells remodel their matrix, and the remodeled matrix reshapes cellular behavior, including which genes are turned on. The reciprocity is physical as much as it is molecular. Alter the matrix, and you alter what the cell can pull on; alter what the cell can pull on, and you alter the internal forces that organize its machinery.

Mechanical cues are unusually persuasive because they reach the cell through the very structures that hold it together. If you zoom in on a cell anchored to its surroundings, the contact points are studded with receptor proteins called *integrins*. Integrins are transmembrane proteins: one end binds to the extracellular matrix outside, the other connects to a dense, multi-protein assembly inside the cell known as a *focal adhesion.* Focal adhesions are not merely rivets. They are busy chemical workplaces filled with enzymes and adaptor proteins, many of which are exquisitely sensitive to tension. In broad terms, a cell tests its environment the way a person tests a railing: by pulling on it. The cell's internal actin filaments and myosin motors (the same partnership that powers muscle contraction) generate contractile forces. If the outside world yields, the molecular structures at adhesions assemble differently than if the outside resists. In that difference lies information.

A concrete, almost unsettling demonstration appeared in the mid-2000s in experiments that asked an apparently innocent question: what does a stem cell "feel"? Mesenchymal stem cells (MSCs) are adult stem cells that can differentiate into lineages such as bone, cartilage, muscle, and fat. Traditionally, biologists coaxed them along particular routes using cocktails of soluble factors. Then Adam Engler and colleagues (2006) cultured MSCs on tunable gels whose stiffness could be set to values similar to soft brain tissue, intermediate muscle, or rigid pre-calcified bone. The cells were given the same genes, the same medium, the same temperature, the same oxygen. The variable was mechanical: how much the surface pushed back.

The result was not a subtle shift but a recognizable reprogramming of fate. On soft, brain-like substrates the cells tended to adopt neuron-like features: a more delicate, branched morphology and gene-expression patterns associated with neural lineages. On stiffer, bone-like substrates they spread, generated higher internal tension, and expressed markers consistent with osteogenic commitment. In the middle, on muscle-like stiffness, they leaned toward myogenic programs. Even more telling, the effect depended on the cell's ability to pull. When nonmuscle myosin II activity was inhibited, the stiffness-driven differences largely collapsed. The mechanical cue was not a passive imprint; it required active probing.

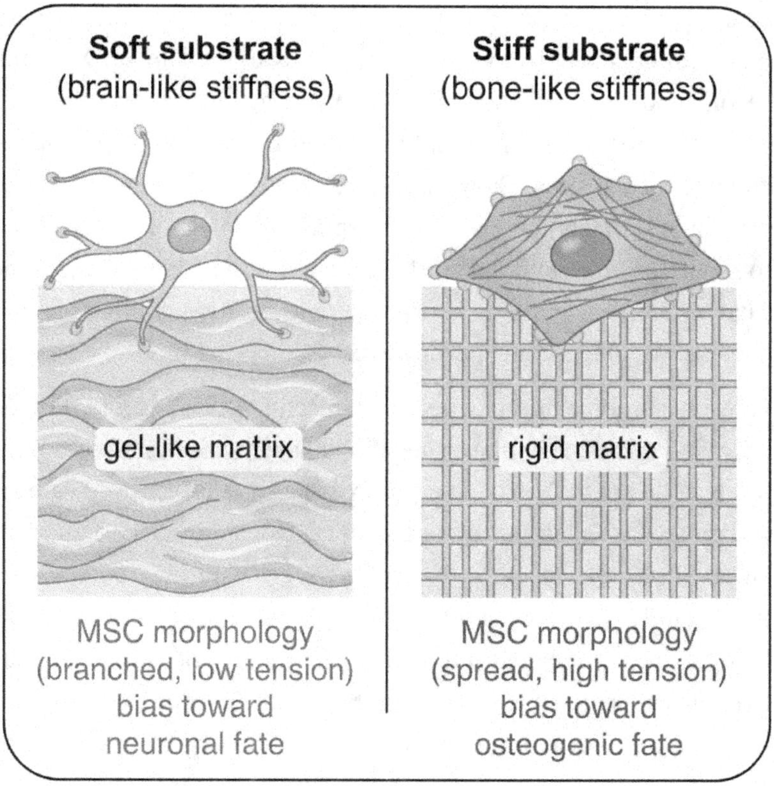


At first glance, this can be mistaken for a laboratory curiosity: gels and petri dishes are not embryos. But the deeper lesson is portable. Tissues differ dramatically in stiffness. Brain is soft; muscle is firmer; cartilage and bone are stiffer still. A stem cell in a healing fracture is not only bathing in growth factors; it is embedded in a changing mechanical landscape as blood clots organize, collagen is laid down, and mineralization begins. Mechanics is one of the ways the body marks a location as "here be bone."

Calling mechanics a *second genetic code* is not meant to diminish genes. It is meant to clarify where biological instructions live. A code, in this sense, is a rule system that maps input to outcome. The genome maps sequences to potential proteins; mechanotransduction maps physical context to patterns of gene activity. Unlike DNA, this second code is not written in letters but in material properties: stiffness, viscosity (how a material flows over time), topography (ridges and pores), and stress (how much force is already present). It is also written in geometry. A cell that can spread widely on a surface tends to generate higher internal tension than a cell forced to remain round. That difference can tilt gene expression, even when stiffness is held constant. Shape becomes a message.

How does a force at the cell surface reach the genome? Part of the answer is mundane engineering. Pull on a tent and the ropes tighten. Many cells are built around a pre-existing internal tension—a baseline tautness sometimes called *prestress*. Donald Ingber popularized an architectural metaphor for this, *tensegrity*, in the 1990s: structures stabilized not by rigid walls but by a balance of tension and compression distributed throughout the framework. In a cell, actin filaments under tension and microtubules resisting compression form a dynamic scaffold. Tugging on one part of the scaffold can, in principle, change mechanical states elsewhere almost instantly, without waiting for molecules to diffuse.

That mechanical continuity matters because some mechanotransduction pathways are fast enough to feel like touch. Stretch an ion channel in the membrane and it can open within milliseconds, letting ions rush in and altering electrical and biochemical states. But other routes are slower and more consequential, because they end at transcription: the process of making RNA from DNA. Here the nucleus is not merely a protected archive. It is a mechanical object with a lamina (a supportive protein mesh) and pores that regulate traffic in and out. Forces transmitted through the cytoskeleton can deform the nucleus, altering chromatin organization and influencing which genomic regions are accessible. A growing body of work also shows that nucleocytoplasmic transport—who gets to enter the nucleus and when—can be mechanically regulated. The most famous examples involve YAP and TAZ, transcriptional co-regulators that tend to accumulate in the nucleus when cells are spread and under tension, and remain more cytoplasmic when cells are soft, round, or mechanically relaxed. In many contexts, YAP/TAZ act like a mechanical throttle on growth and differentiation programs.

This is where the gene-centric picture begins to blur. A gene is not simply "on" or "off" because a chemical switch flipped. It may be accessible because chromatin is reorganized by nuclear forces; it may be activated because a mechanosensitive regulator has entered the nucleus; it may be suppressed because the cell's adhesion machinery cannot assemble the complexes needed to sustain tension. Mechanics is not an accessory variable; it is entangled with what the cell even means by a signal.

A historical anecdote makes the point tangible. Before DNA became biology's central idol, a different sensibility was common: that form emerges from physical law. D'Arcy Thompson's *On Growth and Form* (1917) is packed with arguments that shape is not merely inherited but constrained by mathematics and mechanics. He compared bones to engineering beams, shells to logarithmic spirals, tissues to foams and fluids. For decades, this line of thinking was overshadowed by the triumph of molecular genetics. Thompson's book came to feel like an elegant cabinet of curiosities—beautiful, but not mechanistic in the modern sense.

Mechanobiology revived the spirit of Thompson's question—*why this shape?*—but with molecular handles. The key difference is that modern experiments can tie a physical perturbation to a named pathway and a measurable gene-expression change, and they can do it in living cells, repeatedly, quantitatively. That is how mechanics stops being poetry and becomes causal.

The causal claim is also the controversial one. Skeptics sometimes argue that stiffness is never purely mechanical: change the substrate, and you may also change how proteins adsorb to it; change adhesion, and you change the chemistry of receptor activation. This objection is not a dismissal; it is a warning. Cells do not receive "mechanical" and "chemical" information through separate mail slots. Integrins are both mechanical couplers and biochemical receptors. A focal adhesion is a machine that transmits force and simultaneously recruits kinases and scaffolds. The point of mechanotransduction is precisely that physics and chemistry are yoked. Trying to separate them completely can be like trying to separate the meaning of a sentence from the ink that prints it.

A more productive question is: when does mechanical context become *instructive* rather than merely *permissive*? If a cue is permissive, it allows a program to run but does not choose the program. If it is instructive, it biases the choice among possible programs. Engler's MSC experiments were so striking because the mechanical cue behaved instructively: within a window of time, the stiffness history nudged lineage decisions even when soluble factors were held constant. The cell was not just surviving better on one surface than another; it was adopting different identities.

That word *identity* can sound philosophical, but in biology it has operational meaning. A neuron is not simply a cell with a certain shape. It expresses particular ion channels, neurotransmitter machinery, and cytoskeletal proteins; it forms synapses. An osteoblast is not merely a spread cell. It expresses transcription factors that drive bone matrix production and mineralization. These are stable gene-expression programs reinforced by feedback loops. The scandal—or the wonder—is that mechanical context can help choose which loop becomes self-sustaining.

The notion of a second code also reframes disease. Cancer, for instance, is often introduced as a genetic disorder: mutations that deregulate growth. Yet tumors are also mechanical entities. Many solid tumors are stiffer than the tissues they arise from, in part because of increased collagen deposition and altered crosslinking. A stiffer matrix can amplify cellular tension and promote signaling states associated with proliferation and invasion. In fibrosis, the storyline is explicitly mechanical: injury triggers repair, repair lays down extracellular matrix, the matrix stiffens, stiffness encourages fibroblasts to become more contractile and matrix-producing, and the loop spirals. Genes participate at every step, but the runaway is built from forces as well as transcripts.

Even in everyday physiology, mechanical signaling offers a kind of common sense. Bones thicken under load; muscles grow with resistance training; blood vessels remodel in response to flow. These are organism-scale examples of the same logic seen under the microscope: a living system uses force as information about the world. If the load is persistent, build more structure. If the load is absent, dismantle what is costly to maintain. The genome supplies the tools for both building and dismantling; mechanics helps decide which toolbox to open.

There is an almost intimate implication. A cell is not a passive bead of chemistry. It is an active material that pulls, stiffens, softens, and reorganizes itself in response to resistance. It is constantly asking a question with its cytoskeleton: *where am I, and what kind of world is this?* The answer arrives not as a sentence but as a sensation: the subtle difference between a surface that yields and one that holds. From that sensation, the cell makes long-term commitments, building the very tissues that, at our scale, feel like the solid facts of anatomy.

## 1.2 Spanning the Void: From Nanometers to Meters

A cell feels force in places that are almost unimaginably small. The difference between a relaxed protein and a strained one can be a shift of a few nanometers—a distance so short that, at the scale of a human hair's width, it would be like moving a fingertip across a continent by a few inches. Yet the consequences can be continent-sized in biological terms. A heartbeat, a clenched fist, a swallowed mouthful, the steady lift of the chest during sleep: these are meter-scale motions built from nanometer-scale events repeated, coordinated, and disciplined by architecture.

The surprising part is not that small things add up. A crowd can move a stadium, and a trillion cells can move an arm. The surprise is *how* the information about force survives the scaling. Forces do not simply sum; they get routed, filtered, buffered, and sometimes amplified. A muscle should not twitch every time a single molecular bond breaks. A blood vessel wall should not tear because one cell briefly loses its grip. The body needs a mechanical system that is both sensitive and stable, capable of responding to small, meaningful changes without being derailed by noise. Mechanobiology is, at its core, an account of that system.

The key word is *hierarchy*. Biology is built like a nested set of structures: molecules assemble into complexes; complexes into organelles; organelles into cells; cells into tissues; tissues into organs; organs into bodies. Forces travel up and down this hierarchy. The direction matters. Sometimes an organism-scale force, such as the shear stress of flowing blood, is translated downward into altered gene expression in endothelial cells lining the vessel. Sometimes the flow reverses: a conformational change in a protein in a single cell changes how that cell pulls on the extracellular matrix, and the altered pulling changes tissue shape. The hierarchy is not a ladder with one-way rungs. It is a network with feedback.

At the molecular end of the hierarchy sits a general principle: many proteins are *mechanically switchable*. They do not merely bind partners; they change shape under force. Some proteins have domains that unfold when tugged, exposing hidden binding sites. Others change the lifetime of their bonds when tension rises, a counterintuitive behavior known as a *catch bond*. In ordinary bonds, pulling makes the bond break sooner; in catch bonds, a certain amount of pulling can make the bond last longer, as if the protein prefers to be held. Cells exploit these tricks in adhesions, where stability must be conditional: strong when a cell needs traction, weak when it needs to migrate or remodel.

A focal adhesion can be pictured as a busy mechanical customs office. On the outside, integrins clasp extracellular matrix proteins such as fibronectin and collagen. On the inside, integrins connect to adaptor proteins (talin is a famous one) that link to actin filaments. When actin and myosin generate contractile force, that force is transmitted through talin and other molecules to integrins and into the matrix. Force changes the molecular arrangement of the adhesion; the rearranged adhesion changes biochemical signaling and cytoskeletal organization; the reorganized cytoskeleton changes how force is produced and directed. The loop is self-reinforcing when the environment is stiff enough to resist, and self-limiting when the environment is soft enough to yield.

That last point looks simple—stiff resists, soft yields—but it is the hinge that connects nanometers to meters. The cell is not pulling on a single spring. It is pulling on a composite material made of fibers, crosslinks, water, and living neighbors. Many tissues are *viscoelastic*: they behave partly like a spring (elastic, snapping back) and partly like honey (viscous, flowing). If you pull quickly, they resist more. If you pull slowly and persistently, they may creep and reorganize. A cell probing a viscoelastic matrix must therefore decide not only *how hard* to pull, but *how long* to hold the pull to learn something meaningful about its surroundings. In effect, it performs a mechanical experiment on its own habitat.

Donald Ingber's tensegrity view of the cell, introduced in the 1990s and refined over the following decade, offers a way to make this feel intuitive. In a tensegrity structure, stability comes from pre-existing tension distributed through cables and compression elements. The structure responds globally when pushed locally. You can poke one strut of a well-built tensegrity sculpture and see a change in tension far away. A cell is not a rigid sculpture, but it is pre-stressed: actomyosin contractility establishes a baseline tension that organizes the cytoskeleton. This prestress makes the cell responsive. Without it, a local force would be absorbed like a finger pressing into slack fabric. With prestress, the same force can redistribute, changing the mechanical state of distant components, including those near the nucleus.

A striking example of scale-bridging in the opposite direction comes from the circulatory system. The inside of every blood vessel is lined with endothelial cells, which are bathed in a constant flow of blood. Flow exerts *shear stress*: a tangential drag force that tries to slide the cell surface along the vessel wall. At the scale of the whole body, shear stress depends on heart output, vessel diameter, and branching geometry. At the scale of a cell, it is a continuous mechanical cue that can change cell shape, alignment, and gene expression. Endothelial cells exposed to steady, laminar flow tend to elongate and align with the flow direction, and they upregulate genes that promote a quiescent, anti-inflammatory state. In disturbed flow regions—such as at arterial branches—cells experience fluctuating shear and tend to express genes associated with inflammation and atherosclerosis risk. A meter-scale pattern of blood movement can thus be "read" as a molecular program in a cell. Force becomes fate at the vascular wall, and, over years, this reading can sculpt disease.

The translation from flow to gene expression involves multiple layers. The cell surface contains mechanosensors: ion channels that respond to membrane tension, complexes at cell-cell junctions that feel tugging from neighbors, and focal adhesions that feel traction against the basement membrane. Cytoskeletal tension changes in response, and transcriptional regulators such as YAP/TAZ can shift their localization and activity depending on that tension. The details differ across tissues and conditions, but the logic repeats: a physical input is converted into a change in molecular state, which changes the cell's mechanical output, which changes the physical input.

There is an older, more dramatic historical anecdote that makes the hierarchy vivid: the first time biologists watched living forces shape an embryo with tools built for physics. In the early twentieth century, developmental biology was largely descriptive: careful drawings of embryos at successive stages. The embryo was a marvel, but also a black box. As microscopy improved and quantitative methods entered biology, researchers began to measure forces directly. Techniques like micropipette aspiration (gently sucking on a cell with a glass pipette to infer its mechanical properties) and later atomic force microscopy (pressing a tiny cantilever onto a surface to measure stiffness) made it possible to treat tissues as materials with measurable parameters. The embryo, once a sequence of pictures, became a field of stresses and strains. This shift did not replace genetics; it changed the kinds of questions genetics could answer. If a gene mutation causes a tissue to fold incorrectly, is the mutation changing a chemical signal, or is it changing the tissue's stiffness, adhesion, or contractility? Often it is changing both, and the mechanical side is what makes the visual phenotype inevitable.

At the level of a single cell, the *cortex* is a crucial intermediary. The cortex is a thin, contractile layer of actin and myosin just beneath the cell membrane. It behaves like a tensioned film, a bit like the skin of a balloon, except it is active: it can tighten or relax using chemical energy. A cell's cortex controls its surface tension, its resistance to deformation, and its ability to generate shape changes. When many cells pack together in a tissue, differences in cortical tension and adhesion can drive sorting and rearrangement. This is not metaphorical. It is a mechanical competition between tendencies: adhesion wants surfaces to stick; cortical tension wants surfaces to shrink. The balance can make a tissue behave like an elastic solid in one moment and like a flowing liquid in another.

This balancing act matters because tissue-scale movement is often built from countless microscopic rearrangements. A sheet of cells can elongate, not because each cell stretches like rubber, but because cells exchange neighbors: one interface shrinks, another grows, and the tissue flows. The driving forces may come from polarized contraction of the cortex, oriented cell divisions, or traction against an extracellular matrix. The geometry is managed through local rules, but the effect is large-scale: a narrowing here, a bulge there, a fold that becomes the beginning of a tube. The embryo is full of such controlled flows. What looks like sculpting from the outside is, from the inside, a choreography of tensions.

The mechanical hierarchy also includes the nucleus, not as an isolated command center but as a load-bearing structure embedded in the cytoskeletal web. The nucleus is linked to the cytoskeleton through protein assemblies that span the nuclear envelope, often grouped under the name LINC complexes (for "linker of nucleoskeleton and cytoskeleton"). When a cell pulls on its environment, part of that force can be transmitted through actin cables to the nucleus. The nucleus deforms, sometimes subtly, sometimes dramatically. This deformation can change chromatin organization and influence the movement of regulatory proteins through nuclear pores. A cell's gene expression profile is therefore not only a biochemical response to force; it is also, in part, a mechanical consequence of how force is physically transmitted to the genome's neighborhood.

To see why this matters for body-scale movement, consider muscle. Muscle contraction is the iconic example of nanometers becoming centimeters. At the core of muscle is the sarcomere, a precisely organized array of actin and myosin filaments. Myosin heads bind to actin and, powered by ATP, perform a "power stroke" that shifts them by a few nanometers. That small stroke, repeated by billions of myosin heads in parallel and in series across many sarcomeres, shortens the muscle fiber. The hierarchical organization is essential. If the myosin heads were scattered randomly, their tiny strokes would not add coherently; they would mostly cancel out. Order creates direction.

Muscle also shows how the hierarchy avoids fragility. A single molecular motor can fail without catastrophic effect because there are many motors sharing the load. The tissue is robust, but not blunt. It can be finely controlled because the nervous system can recruit different numbers and types of motor units, and because muscle itself is mechanically tuned: tendons store and release elastic energy; connective tissue transmits forces laterally between fibers; the extracellular matrix influences how fibers slide. Even the sensation of effort arises from mechanosensors in muscles and tendons reporting tension back to the nervous system. The body uses mechanical feedback to coordinate mechanical output.

A less obvious but equally important example is wound healing. When skin is cut, the gap does not close only because cells proliferate and fill in. Cells at the wound edge generate contractile forces, often forming an actin "purse string" that tightens like a drawstring around the wound. Fibroblasts in the underlying tissue pull on collagen, reorganizing it and sometimes causing contraction that can lead to scarring. These are tissue-scale changes driven by cell-scale traction and cytoskeletal contraction. The clinical consequences are deeply human: a contracture after a burn can limit movement for life; an overly stiff scar can ache and itch; a chronic wound can refuse to close because the mechanical and biochemical environments fail to support the necessary traction and remodeling. Healing is a mechanical problem as much as a cellular one.

One reason mechanics is so powerful is that it can act as an integrator, smoothing over the dizzying molecular complexity inside cells. A cell contains thousands of interacting proteins. Trying to predict behavior by tracking every interaction is like trying to predict the motion of a sailboat by cataloguing every molecule of water. Mechanics gives higher-level variables—tension, stiffness, pressure, shear—that summarize many microscopic details. Those variables are not an alternative to molecular biology; they are a bridge between molecular changes and organismal outcomes. When a transcription factor changes expression of a cytoskeletal protein, one consequence is a change in cell stiffness or contractility. That change alters tissue mechanics, which alters force patterns, which feed back into mechanotransduction pathways. The "meaning" of a gene change often depends on where it lands in this feedback web.

There is also a moral, in the old sense of the word: an explanatory lesson. Biology is often described as if it were built from discrete modules, each with a specific function. But mechanical coupling makes the body less modular than we like to imagine. Pull on one part and you may change another. A tumor stiffening its surroundings can influence immune cell behavior. A fibrotic patch in the liver can alter blood flow patterns, changing shear stress elsewhere. A change in posture changes how bones, muscles, and connective tissues load one another, and cells respond to those loads by remodeling. The hierarchy is not just a stack; it is a set of continuous paths along which force travels.

If nanometers can shape meters, it is because living systems have evolved to treat force as both cargo and message. Force is the means by which the body moves, but it is also a signal that tells tissues how to maintain themselves and how to change. The same pull that helps a cell migrate can, if sustained, reprogram gene expression. The same tension that strengthens a tendon can, if excessive or misdirected, contribute to degeneration. In that double life of force—as action and as information—the mechanical hierarchy becomes more than an engineering detail. It becomes a language the body speaks fluently, from the smallest folds of a protein to the sweeping arcs of a limb in motion.

## 1.3 On Growth and Form: A Historical Renaissance

In 1917, a Scottish biologist with a taste for mathematics published a book that still startles readers into a different way of seeing life. D'Arcy Wentworth Thompson's *On Growth and Form* was not a treatise on genes—DNA had not yet been identified as the hereditary material, and "genetic code" was not even a phrase. Thompson looked instead at how shapes arise when matter grows under constraint. He wrote about jellyfish bells and crab claws, about soap films and shells, and about how a change in scale can change the logic of design: a tiny insect can hang from a ceiling with ease, while an elephant must live with thick pillars for legs. To Thompson, anatomy was partly the biography of forces.

What made the book so memorable was not that it denied heredity, but that it refused to grant heredity exclusive rights over form. It insisted that biology is tethered to physics in the most ordinary ways: surfaces carry tension, materials resist deformation, fluids flow, and any growing tissue must negotiate all of these. A bone is not simply "made" by a gene; it is a structure that must bear load. A leaf is not simply "patterned" by a program; it must also catch light and survive wind. Form, in Thompson's hands, looked less like a blueprint and more like a compromise, sometimes elegant and sometimes brute, between what cells attempt and what materials allow.

It is easy to romanticize Thompson as a lone prophet, but his influence also exposes a tension that has shaped biology for a century: the tug-of-war between *description* and *mechanism*. Thompson described and compared, brilliantly, but he could not measure many of the forces he invoked, and he could not name the molecular machinery that might sense or generate them. His work was persuasive as worldview and weak as laboratory recipe. Then came the molecular revolution, and biology learned to do recipes.

The rise of genetics and molecular biology in the mid-twentieth century changed what counted as an explanation. If you could mutate a gene and watch a wing fail to form, you had a causal lever. If you could map a signaling pathway and connect it to a transcription factor, you had a narrative that fit neatly into the cell's chemical vocabulary. Mechanistic success bred a kind of intellectual gravity. Physical constraints did not disappear, but they often became the background: an embryo would "fold" or "invaginate" as if shape change were merely the stage direction appended to a molecular script.

This gene-centered style of thinking was extraordinarily fruitful. It gave us the logic of gene regulation, the discovery of conserved developmental pathways, and a sense of how complex organisms can be built by reusing and repurposing molecular modules. Yet the very success of the molecular approach hid a problem in plain sight:

genes do not act in empty space. They act in tissues with stiffness, tension, and geometry. A gene can specify a protein that binds actin, but whether actin cables pull effectively depends on what the cell is attached to, how its neighbors push back, and what loads the tissue is already carrying. The molecular story is real, but it is not complete on its own.

The historical renaissance of mechanobiology is, in part, the return of Thompson's question under new lighting. The question is not simply "what genes are expressed?" but "what physical situation are those genes expressed *in*, and how does that situation feed back into expression?" The renaissance did not happen because biologists suddenly developed a taste for physics. It happened because the tools arrived: microscopes that can track cell shapes over time in living embryos, materials whose stiffness can be tuned without poisoning cells, and force-measurement techniques that can detect the tiny tractions cells exert on their surroundings.

A second historical figure helps locate the shift: Alan Turing. In 1952, he published a paper proposing that chemical reactions coupled with diffusion could spontaneously generate spatial patterns—stripes, spots, and other motifs—in an initially uniform tissue. Turing's *reaction–diffusion* theory was a physics-first explanation for biological patterning, but its physics was chemical rather than mechanical. It offered a way to get order without an explicit organizer. The fascination of Turing patterns lies in their emergence: complex spatial structure from simple rules.

Turing's paper also reveals how biology began to embrace the idea that physical processes can be generative rather than merely constraining. The embryo might not need a homunculus; it might need instabilities, feedback, and fields. Reaction–diffusion became one pillar of the "physics of development," and mechanobiology became another. Today the more accurate picture is that the pillars are connected: chemical gradients influence contractility and adhesion; mechanical stress influences transport and signaling; and a developing tissue is a multiphysics system in which chemistry and mechanics are coupled.

The mechanobiology renaissance differs from earlier physical thinking in one crucial way: it can name the molecular interfaces where forces become information. Thompson could convincingly argue that shell shapes follow mathematical rules, but he could not point to the cellular apparatus that *implements* the rule. Modern mechanobiology can. The vocabulary is now familiar: integrins at the cell surface; focal adhesions as force-sensitive signaling hubs; the actomyosin cytoskeleton as a contractile engine; the nucleus as a deformable structure whose transport and chromatin organization can respond to stress.

This new vocabulary makes it possible to connect classical observations to modern experiments in a way that feels almost like time travel. Consider one of Thompson's favorite themes: scaling. As an organism grows, forces do not scale politely. Volume increases faster than surface area, and weight grows faster than the strength of many structures unless those structures change shape or material. The adult body solves scaling problems through developmental decisions: thickening bones, reinforcing tendons, reorganizing collagen. Those decisions are not dictated solely by a genetic timetable; they are responsive to load. In a famous example from orthopedics, Julius Wolff in the nineteenth century argued that bone remodels according to the stresses it experiences (Wolff's law). Modern mechanobiology supplies the missing middle: osteocytes and osteoblast-lineage cells sense mechanical strain, translate it via signaling pathways, and adjust the balance of bone formation and resorption. The old observation becomes a cellular mechanism.

A similarly direct link appears in the dynamics of tissue surface tension, an idea that would have delighted Thompson because it borrows from the physics of droplets and foams. When cells stick to each other, and when their cortices contract, a tissue can behave like a viscous liquid over certain timescales. Pascal Lecuit and Pierre-François Lenne have argued that many morphogenetic movements can be understood in terms of effective surface tensions arising from adhesion and cortical actomyosin tension. The language matters: "surface tension" is not a metaphor here but an emergent mechanical property that can be measured and modeled. It provides an explanation for why tissues round up, why cell aggregates can sort like liquids, and why certain boundaries between tissues behave like interfaces. Genes still matter—they control adhesion molecules and contractility—but the tissue-level outcome is more easily predicted by the physics of interfaces than by a list of transcripts.

The same coupling of genes and physics appears in more recent work that brought mechanobiology into popular awareness: the realization that cells respond to the stiffness of the matrix they touch. Once researchers learned to make hydrogels with precisely tunable elasticity and consistent biochemical coatings, it became possible to separate, at least partially, the mechanical feel of a surface from its chemical identity. Dennis Discher and colleagues emphasized that many cells are anchorage-dependent: they require attachment to a solid substrate not just to stay in place but to maintain viability and function. Then experiments such as Engler's (2006) demonstrated that mesenchymal stem cells could be biased toward neuronal, myogenic, or osteogenic fates by substrate stiffness alone. The story was not "genes versus environment"; it was that environment includes a mechanical component that can be as instructive as soluble factors.

A renaissance needs more than good experiments. It needs a community that can talk across disciplines without collapsing into caricature. For a long time, "physics in biology" had two stereotypes.

On one side was the dismissive view: real biology happens at the molecular level, and physics is just common sense. On the other side was the imperial view: biology is applied physics, and genes are merely parameters. Mechanobiology became credible by refusing both. It treats cells as living materials with internal engines, and it treats genes as the means by which those engines are built, tuned, and regulated.

Textbook-level syntheses helped stabilize this middle ground. Works in developmental biophysics, such as the perspective associated with Gabor Forgacs and Stuart Newman, formalized the idea that development can be studied with tools from mechanics, dynamical systems, and materials science without abandoning molecular detail. This matters because it changes what students learn to ask. When a tissue folds, one can ask not only which gene is required, but whether the tissue's elastic modulus changed, whether its viscosity allows flow, whether the pattern of contractility is polarized, whether adhesion is uniform, and how these mechanical fields interact with chemical gradients.

Another ingredient in the renaissance was measurement. It is difficult to overstate how much intellectual history depends on instruments. Once you can measure something, you can argue about it with discipline rather than temperament. In mechanobiology, traction force microscopy made it possible to infer the forces cells exert on compliant substrates by tracking embedded beads. Atomic force microscopy made it possible to map stiffness variations across tissues. Laser ablation in embryos allowed researchers to cut a tiny region of tissue and watch how it recoils, revealing pre-existing tension. Each technique translated an intuitive idea—cells pull, tissues are tense—into numbers that could be compared across labs and models.

This quantitative turn also made mechanobiology clinically legible. Physicians have always palpated tissues: hard lumps, soft edema, taut scars. Those sensations are crude measurements of mechanical state. Mechanobiology offers an explanation for why palpation works. A stiff tumor is not merely a passive consequence of extra collagen; stiffness feeds back into cell behavior, altering migration, proliferation, and immune interactions. Fibrotic tissue is not merely an overabundance of matrix; it is a mechanical environment that can lock cells into a contractile, matrix-producing identity. Even normal aging has a mechanical signature: crosslinking of extracellular matrix proteins can stiffen tissues, changing how cells interpret their surroundings. The Renaissance here is a reunion of the clinician's hand with the molecular biologist's pathway diagram.

There remains a live debate at the center of the field, and it echoes the older tension between genetics and physics: is mechanics a *driver* of form or a *follower*? The truth is that it can be either, depending on the system and timescale. In some cases, mechanical stress patterns arise as a consequence of genetically patterned behavior—cells express different adhesion molecules, sort into domains, and generate tensions at boundaries. In other cases, physical constraints precede and bias gene expression—cells in a stiff region activate different transcriptional programs than identical cells in a soft region. Often the relationship is circular: a slight genetic difference produces a slight mechanical difference, which is amplified by feedback until a distinct form emerges.

The most convincing modern stories are those where the loop is explicitly demonstrated. A morphogen gradient changes actomyosin contractility; altered contractility changes tissue curvature; curvature changes stress distribution; stress changes the localization of mechanosensitive regulators; those regulators feed back on gene expression that further tunes contractility and adhesion. The details vary, but the structure of explanation has become recognizable: coupled fields of chemicals and forces shaping each other in time.

If Thompson could walk through a contemporary mechanobiology lab, he might be amused to see how ordinary his metaphors have become. Engineers and biologists now speak comfortably about stiffness, stress, viscoelasticity, and prestress. They also speak about transcription factors and signaling pathways in the same breath, because it is now clear that molecules and mechanics do not form separate kingdoms. They are different descriptions of the same living process, one emphasizing parts and the other emphasizing interactions and constraints.

There is a quiet philosophical shift hidden in this history. The genome is often treated as an internal cause, the body's true origin, while forces are treated as external disturbances. Mechanobiology complicates that boundary. Many of the forces that matter most are generated by the organism itself. An embryo builds a contractile cortex, pulls on itself, stiffens certain regions, fluidizes others, and thereby creates the mechanical context that guides further development. The organism is both the reader and the author of its mechanical environment.

Thompson once wrote about how a simple coordinate transformation could map one animal's shape onto another, suggesting that form could be understood as a deformation of a mathematical grid. Modern mechanobiology offers a biological version of that idea: the deformations are real, the grid is made of cells and matrix, and the mapping is enacted by molecular machines that can feel resistance. The final shape is not imposed from outside. It is negotiated, continuously, between what genes make possible and what forces make inevitable.

Chapter 2

# Push, Pull, and Flow: The Physics of Being Alive

*Biology is often viewed through the lens of chemistry, yet every living cell inhabits a physical world governed by Newtonian laws. We will examine the invisible architecture of force-from the crushing weight on bone to the rhythmic friction of blood flow-to understand the physical vocabulary cells use to interpret their reality.*

## 2.1 The Architecture of Stress: Tension, Compression, and Shear

A living body is never mechanically quiet. Even in stillness, the lungs tug on the chest wall, the arteries hold pressure like stretched balloons, and muscles keep a low hum of tone so that joints do not collapse into their sockets. To talk about how cells and tissues respond to this constant physical life, a small vocabulary does a surprising amount of work. Three words—*tension*, *compression*, and *shear*—name the main ways forces load biological matter. They are not exotic, and they are not optional: every step, heartbeat, swallow, and breath is some mixture of the three.

Force is a push or a pull. Stress is the more informative quantity: force *per area.* A thumb pressing on skin can apply the same force as a wide palm, but the thumb concentrates it into a smaller area, producing higher stress and a stronger local deformation. Cells do not generally "feel" your intention; they experience stresses and the strains (shape changes) those stresses produce.

Tension is stress that tries to *pull* a material apart. Imagine grabbing both ends of a rubber band and stretching it. The internal molecules resist being separated; the band elongates, and if you keep increasing the load it snaps. Tendons live in tension every time you stand on tiptoe. The Achilles tendon is not merely a rope tied to bone; it is a carefully woven composite whose collagen fibres are arranged to bear repeated tensile stress with minimal damage, yet with just enough give to store and release energy as you run.

Compression is stress that tries to *squash* a material. Sit down and the cartilage in your knee is compressed between femur and tibia; lie on your side and the soft tissues of your hip flatten under your weight. In compression, materials can shorten, buckle, or bulge sideways. The same collagen that is superb in tension behaves differently in compression: it tends to crumple unless supported by other components. Cartilage solves this by combining a tensile collagen network with water-loving proteoglycans that swell like a hydrated gel, creating internal pressure that resists being squeezed. The material is not just "strong"; it is architected for compression.

Shear is stress that tries to *slide* one layer of a material past another. Place your hand flat on a table and push sideways: the friction you feel at the skin is shear. If you hold a deck of cards and push the top card sideways, you create a clean cartoon of shear deformation: the deck becomes a skewed parallelogram. Shear is often less intuitive than pulling and squashing, yet it is everywhere in biology. Skin blisters are as much about shear between layers as they are about pressure. Inside your body, shear also appears when fluids move, dragging on vessel walls, and when tissues experience twisting.

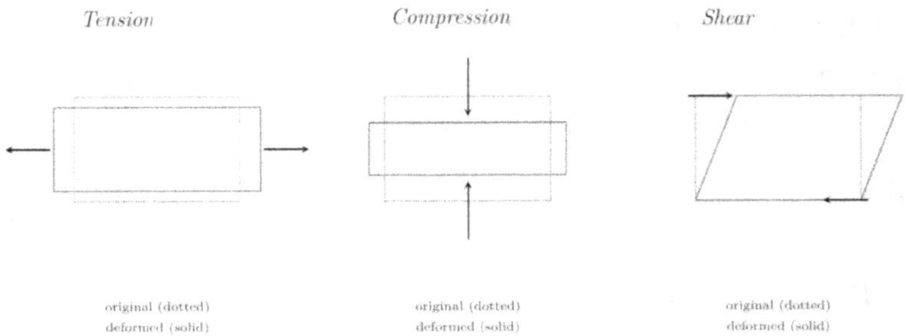


These three loads are easiest to picture with a block of material, but the more important step is to notice them in ordinary life. Pick up a grocery bag: the handle is in tension. Set down a heavy suitcase: the wheels and the floor experience compression. Twist open a jar lid: your skin and the lid experience shear. Biological structures are built from the same physical ingredients as everything else, and so they must answer to the same mechanics. What differs is the cleverness of their designs and the consequences of failure.

A helpful distinction comes from the direction of the stress relative to a surface. Tension and compression are *normal* stresses: they act perpendicular to a surface, either pulling it outward or pushing it inward. Shear stress acts *tangentially*, parallel to a surface, trying to slide it. This matters because tissues are seldom uniform. Skin is layered; arteries have distinct layers with different fibre orientations; even a single cell is a composite of membrane, cortex, cytoskeleton, and nucleus. When layers have different mechanical behaviours, shear is what allows them to move relative to one another—or, when they cannot, what concentrates damage at their interfaces.

The same external action can create different stresses depending on geometry. Consider the bones of the forearm when you do a push-up.

The overall effect feels like compression: body weight travels down through the skeleton. Yet the radius and ulna are not simply being squashed like chalk. They are also bending. In a bending beam, one side is in tension while the opposite side is in compression, with a neutral plane between them. This is why bones are shaped the way they are: their outer layers are placed where tensile and compressive stresses are greatest. The physics is not academic; it is carved into the anatomy.

A famous historical anecdote captures this point with almost theatrical clarity. In the late nineteenth century, Julius Wolff, a German surgeon, argued that bone adapts its internal architecture to the loads it experiences. His idea, now often called Wolff's law, was based on careful observations: the spongy, trabecular structure inside bones seemed to align along directions of stress, like scaffolding placed where it is needed. Modern imaging and biomechanics have refined the details, but the central intuition remains powerful: living tissue is not only stressed; it is *informed* by stress. The femur is not a generic pillar. It is a record of a lifetime of tension, compression, and shear.

That idea becomes more striking when the loads are removed. Astronauts in microgravity lose bone mass; patients immobilized in bed after injury can experience rapid weakening. The body seems to treat underloaded bone as surplus. At the cellular level, osteocytes embedded in bone act as sentinels, responding to mechanical cues and helping direct remodelling by osteoblasts and osteoclasts. The vocabulary of tension and compression, then, is not merely descriptive. It connects everyday forces to the decisions of cells.

To keep the story grounded, return to the knee. When you walk, the joint surfaces press together in compression. But cartilage does not behave like a dry sponge. Much of its resistance to compression comes from water trapped within a charged polymer matrix. Compress it quickly and the fluid cannot escape; the tissue stiffens.

Hold the load and fluid slowly redistributes; the tissue relaxes. Even without naming the mathematics, the message is clear: tissues turn forces into time-dependent deformations. In cartilage, this protects joints from sudden impacts and spreads stresses over a larger area, reducing local damage.

The knee also illustrates why shear deserves its own category. When the femur rolls and slides over the tibia, layers within cartilage can experience shear as their deformation gradients differ with depth. Too much shear, repeated too often, can contribute to cartilage degeneration. In osteoarthritis, pain is not only about "wear and tear" in the naive sense; it involves changes in tissue composition, mechanics, and inflammation. Still, the initiating mechanical language remains the same: where are the tensile, compressive, and shear stresses, and how do they concentrate?

Move from tissues to single cells and the three loads remain relevant, though the scale changes dramatically. A cell attached to its surroundings through adhesion proteins is never a passive blob. It generates internal tension using actin filaments and myosin motors, a process often called contractility. This internal pull places the cytoskeleton under *prestress*—a baseline tension that tunes how the cell responds when external forces arrive. Donald Ingber popularized a vivid architectural metaphor for this in the 1990s with the concept of tensegrity: structures stabilized by a balance of continuous tension and discontinuous compression. In a classic tensegrity sculpture, cables pull continuously while struts resist compression locally, producing a stable, responsive form.

Whether or not one embraces tensegrity as a complete model of the cell, it captures two essential truths. First, forces can be transmitted across a cell faster than biochemical signals diffuse. Tug on the surface at one point and distant structures can shift as stresses redistribute through an interconnected network. Second, the cell's baseline tension changes what it "notices." A slack system can absorb deformation without much change in stress; a taut system translates small deformations into large stress changes. Anyone who has tried to move a loose tent versus a tightly pitched one has felt this difference. Cells, in their own molecular way, live by it.

It is tempting to imagine that mechanical stress is something applied from outside: a weight on a bone, a stretch on a tendon, a pressure on the skin. But biology blurs the boundary between internal and external loads. A healing wound contracts because cells pull on the extracellular matrix, generating tension that closes the gap. A developing embryo folds tissues as cells change shape and exert forces on neighbours. The "environment" of a cell includes not only the outside world but also the forces produced by the cell itself and by its companions. Tension, compression, and shear are shared currencies in this mechanical economy.

To see how these stresses intertwine, consider the simple act of standing. The bones of the leg bear compressive loads, yet the muscles and tendons are in tension as they stabilize joints and prevent sway. In the foot, the plantar fascia is tensioned like a bowstring, supporting the arch. At the contact between foot and ground, shear stresses arise as you subtly adjust balance; too little friction and you slip. If you stand for a long time, tissues in the soles can become sore, partly because sustained compression impedes blood flow and alters the local mechanical environment of cells. A posture is a mechanical arrangement, maintained at energetic cost.

A second concrete example comes from blood vessels, which are often discussed in terms of pressure. Pressure is a normal stress acting equally in all directions within a fluid, and it tends to expand the vessel. That expansion places the vessel wall under circumferential tension, sometimes called hoop stress: the wall must resist the urge to split. But vessels are not merely balloons. When blood moves, it drags along the inner surface of the vessel and creates shear stress at the endothelial lining. Even if the pressure remains the same, changing the flow rate changes the shear. Endothelial cells respond differently to shear than to stretch, and the long-term architecture of vessels reflects both. The mechanical world inside you contains both "push" and "rub."

The subtlety here is that stress is not a thing a tissue *has*; it is a relationship between load and geometry. A thick-walled artery and a thin-walled one can experience the same internal pressure but very different wall stresses. The body exploits this. During chronic high blood pressure, vessels remodel, often thickening their walls, which can reduce the stress experienced by the wall for a given pressure. In engineering, changing geometry to manage stress is standard practice. In physiology, it is adaptation.

Shear also has a reputation for being sneaky because it can arise without obvious shape change. A book on a table is under compression at the contact, but the layers of paper may also experience internal shear if the book is bent or twisted. In tissues, shear frequently concentrates at boundaries: between muscle and tendon, between skin layers, between a cell and its substrate. Many biological failures begin at interfaces because that is where materials of different stiffness and structure meet. Blisters, delamination injuries, certain kinds of vascular damage, and even the detachment of cells from their matrix can all be described in the language of shear and adhesion.

It helps to pair stress with strain. Stress is the cause; strain is the effect: a measure of deformation. Stretch a tendon by a small fraction of its length and you impose tensile strain. Compress cartilage by a percentage of its thickness and you impose compressive strain. Skew a tissue block so that right angles become slanted and you impose shear strain. This relationship is not fixed. Some materials deform easily under small stress; others resist. In living tissues, stiffness is not merely a property; it is a state. Hydration, temperature, biochemical composition, and active cellular tension all modulate the mapping from stress to strain.

An old but enduring figure in biomechanics, Yuan-Cheng Fung, emphasized that living tissues often display nonlinear behaviour: doubling the stress does not necessarily double the strain. Many tissues stiffen as they are stretched because wavy collagen fibres straighten and begin to carry load. This nonlinearity is not a flaw. It allows tissues to be compliant for gentle movements but protective under large loads. A ligament that were stiff from the start would transmit shocks; one that stayed soft would allow joints to dislocate. Nature often chooses a curve rather than a constant.

Yet even that curve is not the whole story. The direction of loading matters because tissues are anisotropic: their properties depend on direction. Wood is stronger along the grain than across it; tendon is stronger along collagen alignment than across it. Skin resists tension differently depending on direction, a fact surgeons use when making incisions along so-called Langer's lines to reduce scarring. Anisotropy is architecture made visible. It reminds us that tension, compression, and shear are not abstract axes; they are invitations to ask, "Along which fibres? Across which layers? Through which junctions?"

At the cellular scale, anisotropy becomes a question of cytoskeletal organization. A cell that spreads on a surface often assembles actin stress fibres aligned along a preferred direction, giving it a directional stiffness. If it is pulled, it may reinforce those fibres, changing its mechanical response. This is why mechanical language matters in mechanobiology: it offers a way to connect a force applied at the outside to a rearrangement on the inside. Tension in an adhesion can become tension in an actin cable; compression on the nucleus can reshape chromatin organization. The details are molecular, but the initiating categories are the same three stresses.

One might worry that reducing biology to tension, compression, and shear is too coarse, like describing music using only loud, soft, and medium. But in mechanics, these are not simplistic labels; they are building blocks. More complicated loads are combinations. Twisting a limb produces shear; bending creates tension on one side and compression on the other; pressure produces tension in a curved wall. Once the basic loads are named, the rich questions follow: where do stresses concentrate, how do structures distribute them, which elements are built for which directions, and how does the living material remodel in response?

The most revealing moments are when the architecture and the stress disagree. When an athlete tears a tendon, it is often not because tension is mysterious but because the loading exceeded the ability to distribute stress safely—sometimes due to fatigue damage, sometimes due to sudden eccentric contraction, sometimes because of a small flaw that concentrated stress like a notch in glass. When a pressure ulcer forms in a bedridden patient, sustained compression and shear deprive tissues of blood flow and mechanically disrupt cells, with damage often beginning deep near bony prominences where stresses concentrate. These are not merely clinical problems; they are demonstrations that stress is not evenly spread. In living tissues, geometry, movement, and material heterogeneity decide where the mechanical bill comes due.

Against that vulnerability stands an equally physical kind of resilience: the body can allocate material where forces demand it. Bone thickens along frequently loaded paths. Tendons adapt to habitual strain. Even at the level of cells, cytoskeletal networks reinforce along lines of tension, and adhesion sites grow under load. The words tension, compression, and shear are therefore more than definitions. They are a way of seeing. They turn the body from a collection of parts into a set of structures carrying loads, distributing stresses, and continuously negotiating with the physics that makes movement possible.

## 2.2 Rivers Within: The Friction of Fluid Dynamics

A hand resting on a tabletop feels solid, certain. The blood inside the hand is the opposite: a moving, slippery medium that refuses to hold shape. Yet that inner river is not a mere delivery service for oxygen and nutrients. Flow itself is information. Cells lining the vasculature read the movement of fluid as a kind of physical language, and they alter the behaviour of the vessel wall accordingly. The crucial word is *shear*—not shear as in twisting cartilage or blistering skin, but shear as *frictional drag* along a surface.

Pressure is the familiar force in plumbing. It pushes outward in all directions. Put a pressure gauge on a pipe and it reports a number without caring whether the water is moving. Shear stress is different. It exists because the fluid moves relative to a surface. In a blood vessel, blood in the middle of the lumen flows fastest; blood near the wall is slowed by friction until, right at the boundary, it is effectively at rest relative to the endothelial surface. The result is a velocity gradient: speed changes rapidly over a tiny distance. That gradient is what creates *wall shear stress*—a tangential drag that tries to sweep the endothelial surface downstream.

This difference between pressure and shear matters because biology does not treat them as interchangeable. A blood vessel can experience high pressure with relatively low shear (think of a stiffened artery where flow is sluggish), or high shear with moderate pressure (a narrow region where flow is fast). Endothelial cells, the thin tiled layer lining every vessel, respond to both, but in distinct ways. They are built to interpret not only "how hard is the blood pushing" but also "how hard is it rubbing."

A simple physical picture helps. Imagine a river flowing over smooth rock. Where the current is fast and steady, the riverbed is scoured clean; where it eddies behind a boulder, sediments settle. The endothelium lives in an analogous situation. In regions of straight, unbranched arteries, flow is often laminar: layered, smooth, and mostly unidirectional. At branch points and curves, the flow can become disturbed, with oscillations and recirculation zones. Decades of vascular biology have linked these patterns to disease: atherosclerotic plaques tend to appear not where pressure is highest, but where flow is complex and shear stress is low or changes direction frequently. Peter Davies drew attention to this relationship in the 1990s by framing shear stress as a central regulator of endothelial phenotype, not a minor perturbation riding on top of chemistry.

The effect is easy to miss because the endothelium is so thin that it looks, under the microscope, like a mere wallpaper. Yet it behaves like an organ: it releases vasoactive molecules that change vessel tone, it controls leukocyte entry into tissues, it regulates clotting, and it remodels the extracellular matrix over long timescales. A single layer of cells, in other words, helps determine whether a vessel stays supple and quiet or becomes inflamed and narrowed. The physical environment of that layer includes the shear stress imposed by flow.

Endothelial cells offer one of the clearest demonstrations that mechanics can act as a signal. Under sustained laminar shear, many endothelial cells elongate and align in the direction of flow, like grasses bent by wind. Their internal actin fibres reorganize, their junctions adjust, and their gene expression changes. Under disturbed flow, they often adopt a less aligned, more irregular state associated with inflammation and permeability. These behaviours are not simply damage responses; they are coordinated programs that depend on mechanotransduction, the conversion of mechanical input into biochemical and genetic output.

A historical pivot in this story came from a shift in what scientists thought they needed to explain. Early physiology could describe the circulation using pressure, resistance, and flow rate, and for a time that seemed like enough. Then pathology forced a more detailed question: why do plaques form in particular places? The distribution is not random. Even within the same artery, one patch of endothelium can be prone to atherogenesis while another, millimetres away, remains relatively protected. Chemical factors such as cholesterol and systemic inflammation matter, but they are distributed widely. Flow patterns are local. The idea that the *pattern of friction* along the wall could influence inflammation helped connect hemodynamics to disease geography.

What does shear stress *look like* to a cell? There is no tiny pressure gauge in the membrane. A cell experiences forces through deformation of structures that couple the fluid-facing surface to the cytoskeleton and nucleus. Several candidate sensors exist, and the truth appears to be plural rather than singular.

One experimentally supported route involves the junctions between endothelial cells. A blood vessel wall is not a sheet of isolated tiles; the tiles are stitched together by adhesion molecules. Work associated with Elaine Tzima and colleagues (2005) described a mechanosensory complex at cell–cell junctions involving PECAM-1, VE-cadherin, and VEGFR2 that can transmit force and trigger downstream signalling pathways, including PI3K activation. The appeal of a junctional sensor is intuitive. Junctions are load-bearing sites where forces on one cell are communicated to its neighbours. A signal that starts at junctions can coordinate a collective response across the endothelial monolayer, helping the vessel behave like a coherent lining rather than a set of independent units.

Another proposed route lies literally in the flow field: the *glycocalyx*, a sugar-rich layer coating the luminal surface of endothelial cells. It is easy to overlook because it can be delicate and hard to preserve in some preparations, but functionally it sits where shear is applied first. Experiments by Florian and colleagues (2007) showed that selectively disrupting components of the glycocalyx could block certain shear-induced responses, notably nitric oxide production, while leaving other outputs less affected. This suggests that endothelial mechanosensing is not a single wire leading to a single switch. It is more like a control panel with multiple inputs: junctional complexes, glycocalyx components, ion channels, and cytoskeletal tension all contributing depending on context. Modern reviews emphasize how glycocalyx components may couple mechanically to membrane and cytoskeletal proteins, offering one plausible path for translating drag into intracellular tension.

If the sensor is plural, the message being sensed is also richer than "fast" or "slow." Shear stress depends not only on flow rate but also on blood viscosity and vessel geometry. Blood is not water. It is a suspension of cells, mainly red blood cells, that deform, align, and migrate within the flow. In large vessels, treating blood as a roughly Newtonian fluid (constant viscosity) can be a decent approximation. In the microcirculation, it can be misleading in ways that matter for biology.

A striking example is the Fåhraeus–Lindqvist effect, described in the early twentieth century and refined through later experiments and theory: as vessel diameter decreases into the range of tens of micrometres, the apparent viscosity of blood can *decrease* rather than increase. This counterintuitive behaviour arises because red blood cells tend to migrate toward the centre of small vessels, leaving a relatively cell-poor plasma layer near the wall. The near-wall region, where shear is applied to endothelium, is therefore not experiencing the same "blood" as the centre. It is experiencing a boundary layer with fewer cells and different mechanical interactions. Pries and Secomb emphasized in their work on microcirculatory rheology (including reviews in the early 2000s) that this is not a small correction; it influences resistance, shear distributions, and the mechanical environment that endothelial cells inhabit.

Even the idea of a smooth velocity profile can break down at these scales. In capillaries comparable to the size of red blood cells, the flow becomes a procession of deforming cells squeezing through narrow tubes. The shear stress experienced by the endothelium is then pulsatile at the level of individual cells and influenced by how those cells tumble and slide. Add bifurcations and the picture becomes more complex: red blood cells do not always split evenly at branch points, leading to phase separation effects and spatial variations in hematocrit. For a cell lining a microvessel, "shear stress" is not a single number; it is a fluctuating physical experience shaped by local traffic.

This has consequences beyond fluid mechanics. Endothelial cells regulate permeability and leukocyte adhesion, and both are sensitive to mechanical cues. Consider inflammation. A white blood cell rolling along a vessel wall is engaged in a mechanical negotiation: adhesive bonds form and break under shear force. Too little shear and cells may stick where they should not; too much and they cannot adhere even when needed. The body exploits this by using flow as a filter: only certain combinations of adhesion molecules and shear allow leukocytes to exit at particular sites. In this way, shear stress participates in the choreography of immune surveillance.

The most emotionally immediate example of shear as signal is not a plaque in an artery but a moment of dizziness. Stand up quickly after lying down and you may feel lightheaded. Part of the reason is gravitational: blood momentarily pools in the legs, reducing cerebral perfusion. The body compensates through reflexes that adjust heart rate and vessel tone. Those reflexes are triggered by stretch sensors in arterial walls (baroreceptors), which respond more to pressure and wall tension than to shear. Yet downstream, within the vascular lining, shear also changes as flow is redistributed. Endothelial release of nitric oxide, a potent vasodilator, is influenced by shear. In a healthy system, these layers of sensing and response help restore stable perfusion. When they fail, as in some cardiovascular diseases, the same physical signals can lead to maladaptive states: stiff vessels, impaired dilation, chronic inflammation.

It is tempting to treat nitric oxide as merely a chemical messenger, but its production can be mechanically regulated. Shear stress can stimulate endothelial nitric oxide synthase activity, promoting nitric oxide release that relaxes smooth muscle and increases vessel diameter. The physics is elegant: increased flow raises shear, which triggers vasodilation, which reduces resistance and helps accommodate the flow. This is not a conscious strategy, but it has the flavour of a feedback system designed by evolution. The vessel uses friction as a proxy for demand.

Flow patterns also guide long-term remodelling. If an artery experiences sustained changes in flow—for example, after an exercise program increases habitual blood supply to muscles—the vessel can remodel its diameter and wall structure. Shear stress appears to play a role in setting the "preferred" calibre: if shear is chronically high, expansion can bring it down; if shear is chronically low, narrowing can bring it up. This principle has limits and interacts with pressure-driven changes, but it captures the idea that vessels are not static pipes. They are living structures that negotiate between competing mechanical demands.

There is an additional subtlety: the endothelium does not experience shear in isolation. The same cells are also stretched circumferentially as the vessel pulses with each heartbeat. The cytoskeleton and adhesion complexes that transmit these forces share components and signalling pathways. A cell must therefore interpret a superposition of mechanical cues with different directions, time scales, and spatial patterns. Disturbed flow is not merely "less shear"; it is often *oscillatory* shear, changing direction over the cardiac cycle. This distinction helps explain why oscillatory, low-mean shear can promote pro-inflammatory gene expression even when the magnitude of shear stress at peaks may not be negligible. The cell seems to care about the *history* of the force: steady wind and buffeting gusts are not the same experience, even if the average speed matches.

One way to appreciate the sophistication of this sensing is to imagine what a purely passive lining would do. A passive coating would simply erode or detach where forces were high. Endothelium does the opposite: it often becomes more aligned and reinforced under steady shear, improving barrier function and reducing turbulence-promoting roughness. The alignment is a physical rearrangement with biological consequences, changing how the cell bears load and how it presents adhesive molecules to passing blood cells. Mechanics and biochemistry become inseparable.

At the scale of whole organisms, fluid shear is not confined to blood. Airflow exerts shear on airway epithelia. Urine flow shears the lining of the urinary tract. Cerebrospinal fluid moves through ventricles and perivascular spaces, potentially providing mechanical cues. Even within tissues, interstitial flow through extracellular matrix can create shear stresses on cells embedded in collagen networks. The magnitudes differ, and the biological roles are less settled in some contexts, but the common theme is that a moving fluid creates directional drag, and cells are exposed to that drag at surfaces and within porous materials.

Mechanobiology often begins with obvious forces: a tendon pulled, a bone compressed. Fluid shear is less obvious because it is distributed, continuous, and invisible. Yet it is among the most constant mechanical signals in the body. Every heartbeat sends a wave of flow that rubs the endothelium, not as damage but as conversation. The vessel wall listens through its junctions, its surface layer, its cytoskeleton, and its nucleus, and it responds by altering tone, permeability, and inflammatory state. The river within does not merely pass through; it shapes the banks as it moves, and the banks, by adapting, guide the river into forms that keep tissues alive.

## 2.3 The Fourth Dimension: Why Tissues Are Liquid Solids

A trampoline feels decisive under your feet: step on it and it pushes back. Honey does not. Press your finger into honey and it yields, slowly and inexorably, then levels itself as if nothing happened. Everyday intuition sorts the world into solids and liquids, and that intuition works just well enough to mislead. Living tissues do not stay obediently on one side of the border. Many behave like solids when you act quickly and like liquids when you wait. Time, in other words, is an extra dimension in the mechanics of life.

This is the practical meaning of *viscoelasticity*. “Viscous” refers to flow, the defining behaviour of a liquid. “Elastic” refers to reversible deformation, the defining behaviour of a spring. A viscoelastic material does both. Pull it and part of the deformation is stored and can be recovered; another part is dissipated as internal friction or rearrangement. If you have ever sat on a memory-foam mattress and then watched the indentation slowly disappear, you have already met viscoelasticity. Your body is filled with materials that behave that way, from skin to ligaments to the cytoplasm inside cells.

It helps to separate two questions that sound similar. First: if a force is applied, how much does a material deform? Second: how *fast* does it deform, and how *fast* does it recover when the force is removed? A steel ruler bent gently springs back almost instantly. A wad of chewing gum stretches, thins, and may not fully return. Tendons lie between: they recoil, but not like steel, and they lose energy with each cycle. That lost energy is why you can feel warmth after repeatedly stretching a stiff joint. The mechanical work you do does not all return to you as motion; some becomes heat.

The most vivid biological demonstration is the difference between a tap and a press. Tap the surface of your skin with a fingertip and it feels firm. Press and hold, and the skin slowly gives way under the same force. A bruise or a welt does not form simply because the stress is high; it depends on the *rate* and *duration*. The same is true in deeper tissues. Cartilage in a knee resists a sudden landing in part because fluid inside it cannot escape quickly, but it deforms more under a prolonged squat as fluid redistributes. A body that had only elastic solids would be brittle and noisy; a body that had only viscous fluids would slump into a puddle. Viscoelasticity allows both support and adaptation.

Engineers have a language for this that is delightfully concrete. An ideal spring stores energy: apply a force and it extends; remove the force and it returns, giving the energy back. An ideal dashpot—a piston moving through a viscous fluid—dissipates energy: it resists motion, but when you stop pushing it does not snap back. Combine spring and dashpot in different ways and you can mimic many behaviours of real materials. The point is not to imagine tiny springs inside a tendon. The point is to recognize that a tissue can carry load *and* relax, resist *and* flow, store energy *and* dissipate it, depending on how long the load persists.

A classic signature of viscoelasticity is *hysteresis*: the path of deformation during loading does not retrace during unloading. Stretch a tendon and then let it recoil, and a plot of stress versus strain would form a loop rather than a single line. The area inside the loop is energy lost per cycle. In practical terms, that loop is why tendons and ligaments are not perfectly efficient springs. Some loss is beneficial, damping oscillations and preventing joints from behaving like overly lively mechanical toys. Some loss is costly, contributing to fatigue and heating under repeated loading. Living systems must tune where on that spectrum they sit.

This tuning is not static. A tissue can change its effective viscoelasticity with hydration, temperature, inflammation, and remodelling. Consider scar tissue. A fresh scar is often stiff, raised, and oddly unyielding. Over months it can soften as collagen is reorganized and cross-links change. The timescale of mechanics becomes part of the clinical experience: stiffness that is obvious at one stage of healing fades at another. The tissue has not merely changed its strength; it has changed the way it relaxes stress over time.

There is a historical reason these ideas became central in biomechanics. In the twentieth century, physiologists and engineers tried to model tissues using the same simple elastic laws used for metals and springs. Those models often failed, not because the measurements were sloppy but because tissues kept revealing a kind of memory. Deformation depended on loading history. A ligament stretched and held at constant length requires less and less force to maintain that length; the stress decays. This is *stress relaxation.* Conversely, hang a weight on a ligament and its length increases over time even if the load stays constant; this is *creep.* These are not obscure laboratory curiosities. They are why braces must be tightened, why posture changes over minutes, and why sustained pressure can become damaging even when it is not extreme.

Yuan-Cheng Fung, one of the founders of modern biomechanics, insisted on taking this seriously. Rather than forcing tissues into a single elastic box, he emphasized constitutive laws that could capture nonlinear, anisotropic, and time-dependent behaviour. The mathematical machinery can become heavy, but the conceptual step is simple: tissues are materials whose response depends on time. Mechanobiology inherits that insight and pushes it inward, asking not only how tissues respond mechanically, but how cells interpret that time dependence as information.

A useful way to make time dependence tangible is to think in terms of *fast* and *slow* processes happening together. A tendon has collagen fibres that bear load quickly when stretched, but it also has ground substance and microstructural rearrangements that take time. A cell has an actin cortex that can tense rapidly, but it also has processes like filament turnover, cross-linker binding and unbinding, and cytoplasmic flow that unfold over seconds to minutes. When you load the material quickly, only the fast processes respond; when you load it slowly, the slow processes have time to participate. The same tissue therefore seems "stiffer" at high rates and "softer" at low rates, even though nothing mystical has happened. You have simply asked different parts of the material to do the work.

The interior of a cell is often described as a gel, and that word captures an important ambiguity: gels can look solid, but they are mostly liquid. Cytoplasm is crowded with proteins, filaments, organelles, and water. When a cell is indented quickly—as in many modern experiments where a tiny probe presses the cell surface—the resistance can be dominated by how quickly water can move through the porous cytoskeletal network. Moeendarbary, Valon, and colleagues (2013) provided evidence that, on short timescales, cytoplasm behaves as a *poroelastic* material: an elastic solid scaffold permeated by fluid, where deformation is limited by fluid redistribution. Poroelasticity is a specific mechanism for viscoelastic-like behaviour. Instead of imagining the cell as a uniform syrup, imagine it as a sponge saturated with water. Press the sponge quickly and it feels stiff because the water cannot escape; press slowly and it yields as water flows through the pores. The same logic applies inside cells, though the "pores" are molecular-scale spaces and the fluid is cytosol.

This poroelastic picture is not the whole story of the cell. Over longer times, cytoskeletal filaments assemble and disassemble, molecular motors generate active tension, and adhesions form and release. Those are not passive relaxations; they are living reorganizations powered by ATP. Still, poroelasticity offers a concrete bridge between the feel of a tissue and the physics of its microscopic structure. It shows how time dependence can arise even without invoking an explicitly viscous cytoskeleton: simply moving fluid through a deforming network takes time.

At the scale of tissues made of many cells, time dependence becomes even more dramatic because cells can rearrange. A single cell can change shape; a group of cells can change neighbours. Under the microscope, an epithelium can look like a mosaic of polygons, stable and neatly packed. Yet over hours, cells exchange positions, divisions insert new tiles, and dying cells are extruded. The tissue can behave like a solid on short timescales and like a flowing material on longer ones. This is the origin of the phrase "liquid solid": the tissue can hold shape and transmit stress, yet it can also remodel and flow without tearing.

Here a different kind of mechanics enters, one that sounds more like surface physics than like springs. When two immiscible liquids meet, their interface has *surface tension*: energy is minimized when the interface is small, so droplets round up. Remarkably, cohesive tissues can show an analogous behaviour, sometimes described as *tissue surface tension*. The idea is not that tissues are literal liquids, but that the balance of cell–cell adhesion and cortical tension can create an effective tendency to minimize interface area between different tissues, leading to rounding and sorting. This concept sits in a long and sometimes contentious lineage, involving hypotheses about differential adhesion and about differences in interfacial tension, but the shared intuition is that mechanics at cell boundaries can produce liquid-like behaviour at tissue scales.

A concrete example makes this less abstract. In classic experiments, mixtures of dissociated embryonic cells can self-organize when allowed to reaggregate: one cell type ends up inside, another outside, reminiscent of oil separating from water. It is hard not to feel a jolt of recognition here. Something about the physical interactions between cells—their adhesions, their cortical contractility, their ability to rearrange—is enough to produce a macroscopic pattern. Developmental biology often emphasizes genes and signals, and rightly so, but mechanics provides a complementary explanation for why certain shapes are natural outcomes of collective cell behaviour. A tissue can "choose" a configuration because it is mechanically favourable, not because any cell has a blueprint of the final geometry.

The liquid-solid duality also shows up in wound healing. Immediately after injury, a tissue must behave solid-like enough to maintain integrity and stop bleeding. Soon after, cells migrate, proliferate, and remodel matrix; the region becomes dynamic, almost fluid in its reorganization. If the tissue remained rigid, it could not close the gap; if it became too fluid, it could not re-establish a barrier. The timing of these behaviours is a mechanical choreography with direct consequences for scarring and function.

There is a temptation to equate "fluid-like" with "soft." They are not the same. A material can be soft but solid, like a rubber eraser. A material can be stiff but able to flow, like very cold pitch, the tar-like substance famous for flowing so slowly that a drip takes years.

The famous pitch-drop experiment, begun in the 1920s by Thomas Parnell and later maintained in Australia, is often used to unsettle our categories: pitch looks solid, but given enough time it flows. Living tissues live in that conceptual space routinely. Bone is mostly solid-like over human timescales, but even bone remodels over months and years. Fascia can feel stubbornly stiff, yet it creeps under sustained stretch. The cornea maintains a precise shape, yet it can slowly change under chronic pressure, as in some forms of glaucoma. The apparent category depends on the timescale at which you ask the question.

Mechanobiology adds a further twist: cells do not merely endure viscoelasticity; they exploit it. A cell pulling on its surroundings can generate a force and then wait to see whether the environment relaxes or resists. A stiff, elastic matrix sustains tension; a compliant or viscoelastic matrix relaxes it. Those differences can feed into signalling pathways that alter gene expression, proliferation, and differentiation. Time dependence becomes part of what the cell "reads" when it reads force. Two environments might have the same instantaneous stiffness, yet if one relaxes quickly and the other holds tension, cells can behave differently. In a world of liquid solids, the question is never only "how hard" but also "for how long."

The same holds for the forces that flow brings to vessel walls. The endothelium is exposed to oscillations with every heartbeat. Some tissues are tuned to respond to pulsatility; others treat it as background noise. The difference is partly biochemical, but it is also mechanical: the cytoskeleton and membrane structures that transmit force have their own time constants. A rapid oscillation may be filtered out, while a sustained change produces a reorganization. This is an underappreciated feature of mechanosensing: a cell is not a perfect detector of force; it is a detector with built-in dynamics.

To keep the picture concrete, return to a simple bodily experience:

stretching a tight hamstring. Pull gently and hold. At first, resistance is high, then it eases. Many people interpret this as the muscle "letting go." In part it is neural, a change in reflex activity. In part it is mechanical: stress relaxation within muscle-tendon units and connective tissue. The easing is real even if you were to stretch a dead tissue sample; biology then layers active responses on top. The subjective experience is the sum of physics and physiology, and it is a reminder that time-dependent mechanics is not an abstract property measured only by instruments. It is part of sensation and movement.

There is also an ethical dimension to getting this right. Medical devices and interventions that interact with tissues must reckon with viscoelasticity. A stent placed in an artery changes local mechanics and flow; the vessel wall adapts over time. A prosthetic joint changes load distributions; bone remodels. Compression garments, braces, and bandages deliver sustained stresses; tissues creep and relax. Even the simple question of how long to hold a pressure dressing after bleeding involves time-dependent tissue behaviour. To treat tissues as purely elastic is to misunderstand how they respond over minutes, hours, and months.

The phrase "liquid solid" might sound like a poetic contradiction, but it is a precise invitation: specify the timescale. On the timescale of a heartbeat, a vessel wall is elastic enough to recoil. On the timescale of days, the same wall can remodel. On the timescale of development, tissues can flow, sort, and fold into organs while remaining coherent. Mechanobiology lives in these shifts of behaviour with time. It asks how structure, fluid movement, and active cellular processes combine to create materials that can be supportive yet adaptable, resilient yet plastic, stable yet capable of changing shape without falling apart.

Chapter 3

# The Inner Scaffold: How Cells Generate Power

*The cell is often imagined as a fragile balloon, yet it withstands mechanical abuse that would tear synthetic materials apart. Beneath the membrane lies a self-repairing architecture of struts and cables that not only resists deformation but actively generates the power to crawl, squeeze, and divide. We uncover the molecular machinery that turns chemical energy into physical force, transforming the cell from a passive passenger into a mechanical agent.*

## 3.1 The Architecture of Resilience

A living cell is soft enough to deform under the gentle drag of flowing fluid, yet tough enough to survive the repeated stretch of a heartbeat, the compression of cartilage, or the abrasive shear of skin against fabric. That combination is not an accident of chemistry. It is engineered, actively and continuously, by a set of internal fibres that behave less like jelly and more like a building whose frame is always under renovation.

Two filament systems do much of the heavy mechanical lifting here: microtubules and intermediate filaments. One behaves like a set of rigid struts that can take compressive load and impose long-range order. The other behaves like a network of extensible ropes that keep the material from ripping when it is pulled, twisted, or struck. Both are woven into the same cytoskeletal fabric as actin, yet their logic is distinct: microtubules are the cell's rapidly reconfigurable beams; intermediate filaments are its resilient safety net.

A useful way to picture this is an architectural idea called tensegrity, introduced into cell biology most famously by Donald Ingber in the 1990s. In tensegrity structures, rigid compression elements do not simply stack on top of each other; they are suspended within a continuous network of tension. The stability comes from balance: struts push, cables pull, and the whole structure holds shape because it is already *pre-stressed.* Cells are not literal tensegrity sculptures, but the metaphor captures something important about why they can be both stable and responsive. Pull on one part of the system and forces redistribute elsewhere, sometimes across surprisingly long distances, because the load is carried by interconnected filaments rather than by a single block of material.

Microtubules are the most obviously beam-like of the cytoskeletal polymers. They are hollow tubes built from tubulin proteins, with a diameter that makes them stout compared with actin filaments, and with a stiffness that is extraordinary for a biological polymer. In mechanical terms, stiffness here means resistance to bending: a microtubule can span micrometres inside the cell without sagging like a rope would. That matters because many cellular problems are problems of geometry. The cell must keep the nucleus away from the leading edge as it migrates. It must position a spindle precisely in the middle when it divides. It must maintain a polarized layout, with different machinery at different ends, even as the outside world pushes and pulls.

Yet microtubules are not permanent girders. The cell can grow them, shrink them, and redeploy them on a timescale of seconds. A microtubule has a built-in asymmetry: one end (the *plus end*) tends to grow faster, the other (the *minus end*) is often anchored near a central organizing site. Growth occurs by adding tubulin units; shrinkage occurs by peeling them away. The striking feature is that microtubules do not simply grow and then slowly erode. They flicker between phases of growth and sudden collapse, a behaviour called *dynamic instability*. If that sounds like a design flaw, it is better understood as a search strategy. A cell can explore space quickly by sending out many microtubule tips, letting some catastrophically depolymerize, stabilizing others when they find the right targets. It is the mechanical analogue of probing a dark room with many canes at once, snapping back the ones that hit nothing.

This restless instability becomes dramatic during mitosis, when the cell divides. Long before anyone knew the molecular details, microscopists could see a strange apparatus form: a spindle-shaped array of fibres that grabs chromosomes and pulls them apart. For much of the twentieth century this spindle was described in almost nautical terms, as if chromosomes were cargo hauled on ropes. The reality is more like a deployable crane. Microtubules assemble into a bipolar machine that can both push and pull. Some microtubules attach to chromosomes through specialized structures called kinetochores; others overlap in the middle of the spindle, where motor proteins slide them against each other; still others brace the spindle against the cell cortex. The spindle does not just *move* chromosomes. It *measures* the cell, finds its centre, and aligns the division plane with the mechanical constraints of the tissue.

A historical anecdote captures how microtubules moved from being vague fibres to being recognized as specific mechanical elements. In the 1960s, researchers discovered that a plant alkaloid, colchicine, could block cell division. It had been used for centuries to treat gout, but in the laboratory it became a tool that made the spindle vanish.

Later, drugs like vinblastine and vincristine, derived from the Madagascar periwinkle, did something similar and became chemotherapies: they preferentially disrupt rapidly dividing cancer cells by interfering with microtubule dynamics. Taxol (paclitaxel) turned out to do the opposite: it stabilizes microtubules so strongly that they cannot reorganize. Both strategies are lethal to the dividing cell, not because microtubules are passive scaffolds, but because their *dynamic* nature is essential. A rigid beam that cannot be reassembled is as useless as a rope that cannot be tied.

This brings us to the other side of resilience: what keeps a cell from tearing. If microtubules are beams, intermediate filaments are ropes and nets. They form a meshwork that is less orderly than microtubules and less famous in introductory accounts, yet they are often what you notice when they fail. The intermediate filament family includes keratins in epithelia (skin and many internal linings), vimentin in many mesenchymal cells, neurofilaments in neurons, and lamins forming a supportive network inside the nucleus. Unlike actin and microtubules, intermediate filaments are not built from a single ubiquitous protein but from many tissue-specific ones. That diversity is a clue to their job: they are tailored to the mechanical life of particular cells.

Mechanically, intermediate filaments have an unusual signature. They are highly extensible: you can stretch them a great deal before they break. They are tough: they absorb energy and resist fracture. A simple way to appreciate the difference is to compare a dry spaghetti strand with a climbing rope. Spaghetti is stiff; it can hold a straight line, but it snaps abruptly when bent too far. A rope is flexible; it cannot hold itself straight, but it will take enormous tension and keep working after repeated abuse. Intermediate filaments behave much more like the rope than the spaghetti. This does not mean they are soft; it means their strength lies in how they respond to being pulled and deformed, especially in large strains where brittle materials fail.

That property becomes life-or-death in tissues that serve as barriers. Skin is constantly subjected to shear: clothes rub, hands grasp, feet slide in shoes. Epithelial sheets must stay intact under these insults. Keratin intermediate filaments form a dense network in epithelial cells and connect to specialized junctions that stitch cells to each other (desmosomes) and to the underlying basement membrane (hemidesmosomes). The result is not merely that each cell is reinforced, but that the entire sheet becomes a mechanically integrated fabric. Load applied to one patch can be distributed across many cells, reducing the chance of a local tear propagating into a blister.

When this system breaks, the consequences are immediate and visible. A group of inherited disorders known as epidermolysis bullosa causes skin to blister from minor friction, sometimes from something as gentle as being lifted. Several forms of the disease arise from mutations in keratins or in the junctional proteins that connect keratin networks between cells. The molecular details differ among subtypes, but the mechanical logic is the same: remove the tough, energy-absorbing ropes, or detach them from the seams, and the tissue behaves like a poorly stitched fabric. It may look intact at rest, but it fails catastrophically under everyday stress.

The same logic shows up in the nervous system, though the failure mode looks different. Neurons have long, thin axons that can span centimetres to metres. These processes must be mechanically supported, not only against external deformation but against their own internal logistics: the transport of vesicles and organelles along microtubule tracks. Neurofilaments, a type of intermediate filament, contribute to axonal calibre and mechanical stability. In traumatic brain injury, axons can undergo rapid stretch; one proposed contributor to damage is the disruption of cytoskeletal architecture under strain. Here, too, resilience is not a luxury. It is a precondition for function.

An intermediate filament network is often described as more stable than actin, less dynamic than microtubules. That is broadly true, but it can be misleading if it suggests passivity. Intermediate filaments constantly exchange subunits and reorganize, just on different timescales and under different controls. Their assembly is built from coiled-coil dimers that form staggered tetramers and then pack into rope-like filaments. Many of the control knobs are chemical: phosphorylation can loosen networks and change their interactions. But the most dramatic control knob is mechanical: intermediate filaments respond to load by changing their own internal conformation, uncoiling and sliding in ways that dissipate energy. In other words, they are not only *bearers* of tension; they are *processors* of strain.

Microtubules and intermediate filaments do not operate in isolation. Microtubules can resist compression, but slender struts have a well-known failure mode: buckling. Push hard enough on a beam and it bows sideways. In cells, microtubule buckling is not rare; it can be observed when cells contract strongly or when microtubules encounter obstacles. Buckling is not simply a sign of collapse. It is a way to limit compressive forces and prevent breakage, and it can be used as a mechanical signal. But buckling also illustrates why a beam needs a surrounding tension network. A strut embedded in a mesh is stabilized laterally; it can take more load without folding. Conversely, a tension network without struts can collapse inward under its own tension, like a tent without poles. Resilience emerges from the partnership.

This partnership is partly physical and partly organized by specific linker proteins. Microtubules are crosslinked to other cytoskeletal systems; intermediate filaments are connected to cell junctions and to the nucleus. The nucleus itself is not a passive passenger: it is mechanically reinforced by lamin intermediate filaments lining its inner membrane. That reinforcement matters because the nucleus is both fragile and vital. Excessive nuclear deformation can rupture the nuclear envelope, mislocalize DNA repair factors, and threaten genome integrity. Cells migrating through tight spaces in tissues must squeeze the nucleus; the lamin network is a critical determinant of whether that squeeze is survivable. Here resilience touches disease in another way: lamin mutations can cause muscular dystrophies and premature aging syndromes, conditions where mechanically stressed tissues and long-term maintenance are central.

The cell's internal architecture also has an economic dimension. Building and maintaining a stiff microtubule array costs energy and resources, but it buys organizational power: directed transport, rapid spatial exploration, a scaffold for polarity. Building intermediate filament networks buys mechanical insurance: an ability to withstand rare but damaging events, the biological equivalent of a crumple zone. Different cell types make different choices because they live different lives. A white blood cell squeezing through tissues prioritizes actin-driven shape change and may keep intermediate filaments relatively sparse. A skin cell at the body surface invests heavily in keratins because it expects shear. A muscle cell invests in specialized contractile architecture but still requires intermediate filament-like systems to transmit force and prevent tearing.

There is also a subtlety in how cells distribute resilience across scales. At the level of a single filament, microtubules are stiff and intermediate filaments are tough. At the level of networks, the picture depends on geometry and connectivity. A sparse microtubule array can behave like a set of antennae, probing space but not strongly reinforcing the whole cell. A dense, crosslinked array can act like a cage. Intermediate filament networks, because of their extensibility, can allow large deformations while maintaining connectivity, which is exactly what tissues often need. A sheet of epithelium must stretch and recoil as organs move, yet it cannot afford microscopic cracks that would become macroscopic wounds.

One of the most revealing mechanical experiments on cells involves pulling, not pushing. If you attach a tiny bead to the surface of a cell and pull with a calibrated force, the cell deforms in a way that reflects not just the local material but the connectivity of the whole cytoskeleton. Cells often stiffen when stretched, a property called *strain stiffening*. Intermediate filaments contribute strongly to this. Strain stiffening is the reason a climbing rope feels soft when gently tugged but becomes unyielding when you load it with a body. In a cell or tissue, strain stiffening is a safety feature: small deformations remain easy (useful for normal physiology), but large deformations trigger a sharp rise in resistance (useful to prevent damage). It is one of the clearest examples of how biology exploits nonlinear mechanics rather than fighting it.

Microtubules, too, have nonlinear stories. Their dynamic instability means that mechanical load can stabilize or destabilize them depending on context. A microtubule pushing against the cell cortex can slow its growth and increase the probability of catastrophe, effectively coupling mechanics to architecture. Motor proteins that walk along microtubules can generate sliding forces, and the microtubules can respond by bending, buckling, or aligning. These interactions turn microtubules into more than passive tracks. They are parts of a feedback loop where force changes structure and structure changes the pattern of force.

The most satisfying picture is not of three separate cytoskeletal systems with separate jobs, but of a composite material whose properties can be tuned. Microtubules give the cell long-range order and compressive support, but they rely on a tense network to avoid buckling and to transmit forces effectively. Intermediate filaments give the cell and tissue toughness, but they rely on anchor points and junctions to distribute load. Both coexist with actin-based systems that generate and bear tension rapidly. Together, they allow cells to behave like materials that engineers still struggle to build: structures that are strong yet lightweight, stiff in some directions yet flexible in others, robust to damage yet capable of self-repair.

A final image makes the point concrete. Imagine an epithelial cell in your skin as you rub your thumb against your palm. At the surface, shear forces try to slide cells past each other. Inside each cell, the keratin network tightens and shares the load with junctions that link neighbour to neighbour. Microtubules, reorganizing in the background, keep intracellular transport running and maintain polarity so the cell can keep renewing its surface. Nothing about this is static, and nothing about it is merely supportive. The architecture is alive: beams that can be dismantled and rebuilt, ropes that harden when pulled, and a fabric that turns mechanical stress into survival rather than rupture.

## 3.2 The Molecular Engines of Force

A cell can be mechanically resilient and still be helpless. Resilience keeps form from falling apart; force changes form on purpose. When a white blood cell squeezes between endothelial cells to reach an infected tissue, when the edge of a wound draws closed, when an embryo folds and pinches to carve out organs, something inside the cell must *do work*: pull against resistance, spend energy, and create motion that is biased rather than random.

The central machinery for that work is built from actin and myosin. Actin is a protein that can assemble into filaments, thin cables that can be arranged into meshes, bundles, rings, and arcs. Myosin is a motor protein, an enzyme that turns the chemical energy of ATP (adenosine triphosphate) into a directed mechanical change in shape.

Put them together in the right geometry and the cell acquires a kind of internal muscle. It is not a miniature version of a biceps, neatly striated and regular. It is more like a contractile fabric: a living material that can pull from many points at once, thicken where needed, soften elsewhere, and constantly rebuild itself.

The first surprise is that cells generate force in two very different ways using the same actin building blocks. One is pushing: actin filaments can grow at one end and, if they grow against a membrane, they exert a protrusive force. This is how many cells make flat, ruffling leading edges. The other is pulling: myosin motors can slide actin filaments past one another to produce contractile tension. Pushing and pulling are not in competition; they often operate in the same cell at the same time. A crawling cell, for example, pushes its membrane forward by actin assembly at the front and pulls its rear forward by actomyosin contraction, while gripping the environment through adhesions.

Contractile force is the more subtle of the two, because a motor sliding along a filament sounds like a straightforward piece of engineering until you ask a simple question: how does sliding become *shortening*? If two actin filaments are not arranged in an orderly way, a motor might just shuffle them around without producing net contraction. The cell's solution is statistical and architectural. It does not require every filament to be aligned; it requires enough connectivity, enough polarity bias in local arrangements, and enough nonlinear mechanics that tension can be focused and transmitted. In living cells, geometry and chemistry conspire to make a disordered mesh behave, for practical purposes, like a coordinated engine.

Myosin itself comes in many flavours. The best-known is the myosin of muscle, which forms the thick filaments in sarcomeres. Cells that are not muscle mostly use *non-muscle myosin II*, a close cousin adapted for more flexible, rapidly remodelled structures. Like its muscle counterpart, myosin II forms bipolar filaments: assemblies with motor heads at both ends. That bipolarity is crucial. A single myosin head walking on actin would be a transporter, like a person moving along a rope. A bipolar assembly is more like two groups of people pulling on two ropes in opposite directions, drawing them together. If each group walks toward the middle, the ropes slide inward and the distance between their far ends shrinks. In a cell, those "ropes" are actin filaments; the "people" are myosin heads; the "walking" is powered by ATP.

ATP matters because it is the cell's rechargeable battery. Hydrolysing ATP (splitting it into ADP and phosphate) releases energy that can be harnessed by enzymes. Myosin is one such enzyme. It binds ATP, changes its shape, binds actin, releases phosphate and ADP, and in the process performs a mechanical step. The elegance is that chemical state and mechanical state are coupled so tightly that you can read one from the other. When ATP binds, myosin's grip on actin loosens; when phosphate is released, the grip tightens and a force-producing movement occurs. This coupling is sometimes called a mechanochemical cycle: chemistry biases mechanics in one direction, preventing the motor from simply jittering back and forth like a thermal toy.

A single myosin head is a molecular lever. It has a motor domain that binds actin and handles ATP, and a "neck" or lever arm that amplifies small changes in the motor domain into a larger swing. The familiar diagram shows the myosin head "rowing" along actin, and that is not far from the truth, provided one remembers that the oar is a protein arm only a few nanometres long and the water is the crowded interior of the cell. What matters for the cell is not the romance of the analogy but the outcome: a biased step that can generate force against a load.

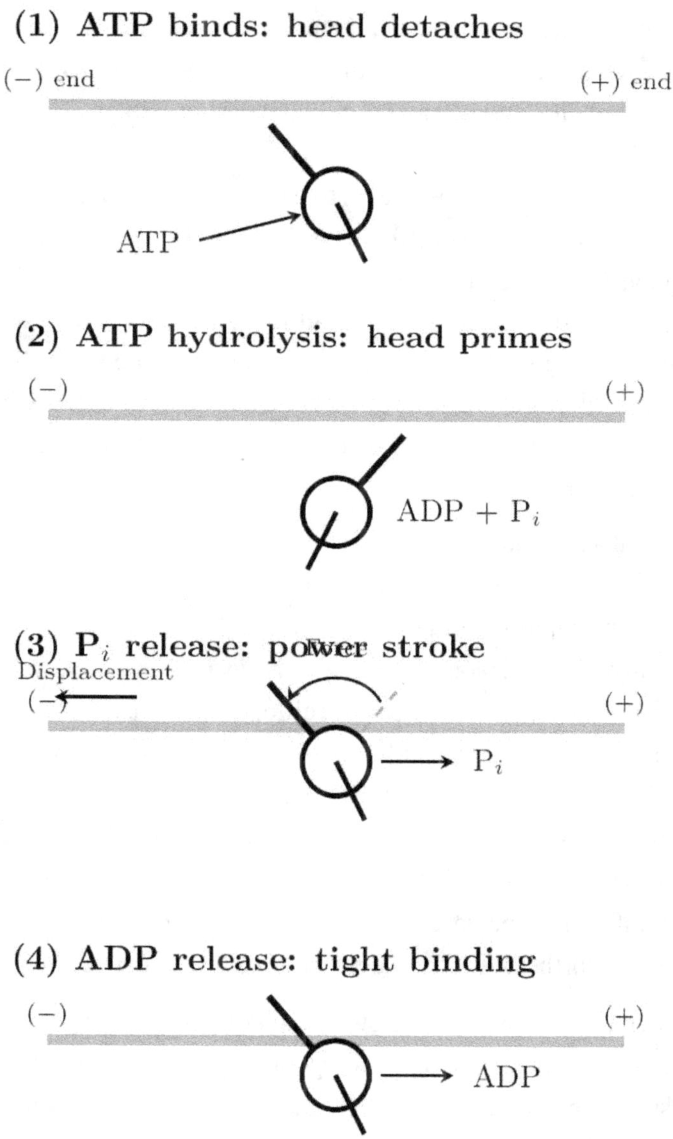


In muscle, myosin and actin are arranged with exquisite regularity, and the resulting contraction is clean and powerful. In most cells, the arrangement is more improvisational. Consider a fibroblast, the kind of cell that crawls into a wound and helps rebuild connective tissue. Under a microscope, fibroblasts show thick actin bundles called stress fibres that span the cell like tension cables. These fibres are anchored at their ends to focal adhesions, molecular rivets that connect the internal cytoskeleton to the extracellular matrix. Myosin II filaments intercalate into these bundles, pulling on them and generating tension. That tension is not wasted: it is fed into the adhesions and into the surrounding matrix, allowing the cell to tug on its environment, align collagen fibres, and even "test" the stiffness of the substrate. A stiffer matrix allows higher tension to build, which in turn promotes larger, more mature adhesions, a feedback that can steer cell behaviour.

The feedback is not merely a convenient addition; it is part of what makes force generation useful. If a cell could generate tension but could not couple it to the outside world, it would simply contract internally like a person clenching a fist in midair. Adhesions give traction. They also turn mechanical load into biochemical signals, altering gene expression and fate decisions. Cells in a soft environment often behave differently from cells on a stiff one, even when the chemical cues are identical. This mechanosensitivity is one reason scar tissue, which is stiffer than healthy tissue, can be biologically self-reinforcing: increased stiffness promotes increased contractility, which can further remodel the matrix into a stiffer state.

A concrete example shows how physical force can shape a clinical outcome. In wound healing, a temporary "purse string" of actin and myosin can form around the edge of a wound, particularly in epithelial sheets. Imagine a circular tear in a fabric. One way to close it is to patch it; another is to tighten a drawstring around its rim. Cells often use the drawstring strategy. Actin filaments align around the wound edge; myosin II generates contractile tension; the ring tightens and the wound area shrinks. This is not just a neat trick. Excessive or prolonged contractility can contribute to contractures and pathological scarring. Too little contractility can lead to chronic wounds. The balance involves not only growth factors and inflammation but also the mechanical state of the cells and their matrix.

The story of myosin as a motor also has a historical turning point. For decades, the sliding filament model of muscle contraction, developed in the 1950s, framed myosin as the central engine of movement. But the broader cell biology of force took longer to crystallize, because non-muscle cells lacked the obvious striations of muscle and because their motility seemed too variable to reduce to a simple machine. The realization that non-muscle myosin II forms bipolar filaments and that its activity is regulated by phosphorylation of its light chains gave cell biologists a handle: cells could tune contractility up and down with signalling pathways, assembling and disassembling contractile units as needed rather than maintaining a permanent apparatus. Miguel Vicente-Manzanares and colleagues emphasized in the late 2000s how central non-muscle myosin II is to adhesion and migration, not as a brute force generator but as a coordinator that couples mechanics to signalling.

Still, there is a puzzle that matters for mechanobiology: how does a disordered actin mesh produce a net contractile stress? A tidy sarcomere works because filament polarities are arranged so that myosin pulls toward the centre. In a random network, half the time a myosin filament might be expected to pull filaments in ways that do not shorten the network. One reason contraction nevertheless emerges is that actin networks are mechanically nonlinear. Under compression, actin filaments can buckle; under tension, they resist strongly. If myosin pulls two filaments together, one part of the network may go into compression and buckle easily, while another part goes into tension and transmits force efficiently. Compression becomes mechanically "cheap" and tension becomes mechanically "expensive," and the net effect can be a collapse toward a contracted state. Michael Murrell and colleagues highlighted in the 2010s how such nonlinear mechanics, together with network connectivity and remodelling, can allow large strains in non-sarcomeric actomyosin systems.

Connectivity is the other key. Actin filaments are crosslinked by many proteins, some of which act like staples holding filaments at angles, others like zipper teeth holding them in parallel bundles. If the network is too loosely connected, myosin activity simply stirs filaments without transmitting long-range stress. If it is too tightly crosslinked, the network can become jammed, resisting rearrangement and sometimes preventing effective contraction. Cells tune this continuously. They can rapidly assemble branched networks at the leading edge for pushing, then reorganize them into bundles and arcs for pulling. They can localize myosin activity to particular regions, creating contractile gradients that polarize the cell.

The interplay of actin assembly and myosin contraction is especially vivid in cytokinesis, the final step of cell division when one cell pinches into two. The mitotic spindle positions the plane of division, but the physical pinching is performed by a contractile ring: actin filaments and myosin II arranged around the cell equator. As myosin pulls on the actin network, the ring tightens, the membrane invaginates, and a cleavage furrow deepens. Unlike a rigid noose, this ring must keep renewing itself as it constricts. The circumference decreases, components turn over, and the cell must coordinate mechanics with membrane trafficking to add surface area where needed. It is a small-scale feat of engineering that repeats countless times during development and tissue renewal.

Actomyosin contractility also underlies a less obvious kind of motion: flows within the cortex, the thin actin-rich layer under the plasma membrane. If myosin activity is higher on one side of a cell, that side can contract more strongly, generating cortical flows that transport proteins and reorganize polarity. In embryos, such flows can help establish body axes. In single cells, they can help cluster receptors or drive shape changes. The remarkable thing is that the cortex behaves like a material that is neither solid nor liquid but something in between: a viscoelastic gel that can support tension over short times yet remodel over longer ones. That combination lets the cell maintain shape while still being able to change it.

Because myosin is an ATPase, contractility has a metabolic price. Cells that are starved of ATP soften and lose tension; their edges sag; their internal organization degrades. This dependence turns mechanics into a readout of cell health. It also raises a more subtle point: force generation is not simply a matter of turning motors on. The cell must decide where to spend energy. Myosin filaments assemble where tension is needed, and disassemble when it is not. Regulatory pathways adjust myosin activity and localization. The cell's mechanical state becomes a dynamic budget, allocating contractile effort to migration, division, maintaining junctions, or resisting deformation.

There is a temptation to imagine that myosin always produces "good" tension, as if more force were inherently better. In disease, the opposite can be true. In asthma, airway smooth muscle can become hypercontractile. In fibrosis, excessive contractility of fibroblasts and related cells can stiffen tissues and distort organ architecture. In cancer, altered actomyosin tension can change how cells migrate and invade, and can influence how they sense and respond to the stiffness of their surroundings. These are not just chemical diseases with mechanical side effects; they are often mechanical diseases with chemical signatures.

The actin—myosin engine is also not the only motorized system in the cell. Microtubules have their own ATP-powered motors, kinesins and dyneins, which haul cargo over long distances and help organize structures like the mitotic spindle. The discovery of kinesin in the 1980s was a reminder that cellular power is distributed: different tracks and different motors are specialized for different scales of transport and force. But for rapid shape change, for gripping and pulling, for the tactile conversation between a cell and its surroundings, actomyosin remains the workhorse.

Watching a cell under the microscope, one sees the effects of this engine as a kind of purposeful restlessness: edges protrude and retract, fibres thicken and thin, adhesions appear, mature, and dissolve. Underneath is the same basic cycle repeated across billions of motors: bind, pull, release, reset. The miracle is not any single stroke. It is the coordination that emerges when countless tiny levers pull on a shared, remodelled network, turning chemical fuel into the tension that holds a cell together while it changes its mind about where to go.

## 3.3 The Living Skin

A cell's boundary looks simple in cartoons: a neat oval line separating inside from outside. In reality it is a thin film only a few nanometres thick, closer to a soap bubble than a wall. The plasma membrane is made mainly of lipids, molecules with water-loving heads and oil-loving tails that assemble into a bilayer. This bilayer is an astonishing invention: it keeps salts on one side and proteins on the other; it hosts receptors and ion channels; it can bend into tubes and vesicles; it can fuse and split. Mechanically, though, it is fragile. A pure lipid bilayer tears easily if pulled too hard, and it cannot resist compression without wrinkling. The reason cells can be squeezed, stretched, and poked without instantly rupturing is that the membrane is not acting alone. Just beneath it lies an active, contractile meshwork of actin and associated proteins: the cortex.

The cortex is sometimes described as a shell, but that suggests something rigid and inert. A better mental picture is a woven fabric stretched under the membrane, continually repaired even as it is tugged. It is thin (typically a fraction of a micrometre), dense, and dynamic. Actin filaments in the cortex turn over rapidly; crosslinkers bind and unbind; myosin motors pull on the network and generate tension. The membrane and cortex are mechanically coupled, so that they share load. The membrane contributes a kind of surface tension, and the cortex contributes an actively maintained contractile tension. Together they form the cell's *living skin*: a composite boundary that is both a barrier and a sensing surface, both a mechanical container and a responsive interface.

Surface tension is a familiar idea from everyday life. It is why a water droplet beads up, why a water strider can stand on a pond, why a soap film snaps taut across a wire frame. In physics, surface tension is an energy cost per unit area: stretching a surface requires work, so the surface resists expansion. The plasma membrane behaves as if it has a tension, but the origin is more complex than in a simple fluid. Part of it comes from the lipid bilayer itself, which resists stretching because lipids do not like to be pulled apart. Part comes from the membrane's attachment to the cortex: if the membrane is glued to a tense network, pulling the membrane also pulls against those attachments. Cell biophysicists often talk about an *effective* membrane tension that bundles these contributions together because, for many experiments, what matters is how hard it is to pull out more membrane area.

Here is where a basic law from nineteenth-century physics becomes unexpectedly relevant. Laplace's law relates the pressure difference across a curved surface to its tension and curvature. For a simple spherical surface, it can be written as

$$\Delta P = \frac{2T}{R},$$

where $\Delta P$ is the pressure difference between inside and outside, $T$ is the surface tension, and $R$ is the radius. The smaller the radius, the larger the pressure needed to hold the shape at a given tension. In cells, this relationship quietly governs many behaviours. If cortical tension rises while cell volume stays roughly constant, the cell tends to round up. If pressure rises locally or the cortex weakens, a bulge can form. Even without writing equations, one can feel the intuition: a taut skin over a pressurized interior will smooth itself into a rounded shape.

Cells exploit this interplay in dramatic ways. During mitosis, many animal cells become rounder. Part of that rounding comes from changes in adhesion to the substrate, but much of it reflects increased cortical tension and altered pressure regulation. Rounding is not cosmetic. It creates space for the mitotic spindle to form and function, and it provides a more symmetric geometry for division. In a crowded tissue, rounding can be mechanically resisted by neighbours, and that resistance can affect the fidelity of division. The *living skin* is therefore not only an envelope; it is a machine component in one of the cell's most consequential operations.

The coupling between membrane and cortex is not automatic. It is mediated by specific proteins that act as adjustable connectors. A prominent family is the ERM proteins—ezrin, radixin, and moesin—which can bind membrane components (often via specific lipids such as PIP_2) and also bind F-actin. They are regulated by conformational switching and phosphorylation, which means the cell can tune how tightly the membrane is tied to the underlying mesh. Tight attachment makes the surface behave more like a reinforced laminate; looser attachment allows the membrane to slide, deform, and locally detach.

That local detachment produces one of the strangest and most revealing forms of cell motion: blebbing. A bleb is a spherical bulge of membrane that balloons outward from the cell surface, often forming in seconds. At first glance a bleb looks like a purposeful protrusion, like a tiny arm. But its initial expansion does not require actin polymerization pushing at the edge. It is driven primarily by pressure. If a patch of membrane detaches from the cortex or if the cortex locally ruptures, the pressurized cytoplasm can push the membrane outward, inflating a bleb like a weak spot in a bicycle tire. Only later does actin reassemble under the bleb membrane, forming a new cortex that can stabilize the bulge and, with myosin activity, retract it.

Blebs were once treated as pathological curiosities, associated with dying cells or cells damaged by toxins. Work by Ewa Paluch, Gareth Charras, and others in the 2000s helped reframe them as a legitimate mode of migration and morphogenesis. In confined environments—for example, when immune cells move through dense tissue or when cancer cells squeeze through narrow gaps—bleb-based motility can be effective. The cell uses pressure-driven protrusions to explore and push forward where actin-driven lamellipodia would be hindered. The *living skin* is doing two things at once: it is a pressure vessel and a shape engine, able to switch between modes depending on confinement, adhesion, and the coupling strength between membrane and cortex.

This switch reveals something that can be easy to forget: cells are pressurized. The inside of a cell is crowded with proteins and polymers, and ion gradients pull water in. Cells control their volume by pumping ions and by regulating channels, but at any moment there is typically an internal pressure relative to the outside. That pressure does not automatically create protrusions; it is contained by the cortex and the membrane. Myosin-generated cortical tension can increase internal pressure by squeezing the cytoplasm, much as tightening a belt increases the pressure in a soft bag. Conversely, if pressure rises without a matching increase in cortical support, the surface can bulge and bleb. Pressure and tension are not rivals; they are partners in a constrained negotiation.

A concrete experimental trick makes this negotiation visible. In so-called tether pulling experiments, researchers attach a microscopic bead to the cell membrane and then pull it away using an optical trap or a microneedle. The membrane draws out into a thin tube, a tether. The force required to hold the tether reflects both the membrane's bending stiffness and the effective membrane tension, including the contribution from membrane-cortex attachments. Michael Sheetz and Jianwu Dai argued in the 1990s that membrane tension can act as a control parameter for cell behaviour, influencing processes like endocytosis and secretion. The intuition is again familiar: if the skin is taut, adding more surface area is hard and the cell may respond by delivering membrane through exocytosis; if the skin is slack, the cell can afford to internalize membrane. The membrane is not merely a passive recipient of traffic; its mechanical state can bias the traffic itself.

Whether membrane tension is a global variable or a patchwork of local states remains an active debate. In a simple soap film, tension equilibrates quickly across the surface. In a cell, the membrane is studded with proteins, connected to a cortex, partitioned into domains, and constantly being reshaped by trafficking. Local tension gradients can exist, and proteins can sense and respond to them. Yet cells also show behaviours that suggest some degree of global coupling, as if tension changes in one region can influence distant regions. The truth is likely conditional: sometimes tension propagates quickly, and sometimes it is trapped locally by friction, obstacles, and differential attachment. The *living skin* is not a uniform sheet; it is a busy, heterogeneous landscape.

Mechanosensing at the membrane adds another layer. Some proteins in the membrane change conformation when the bilayer is stretched or curved, and ion channels are among the most direct examples. Piezo channels, discovered in the 2010s, are large mechanosensitive channels that open in response to mechanical stimuli. Their structure suggests a design tuned to membrane deformation: a curved architecture that can respond to tension and curvature changes by gating a pore. When Piezo channels open, ions flow, electrical and chemical signals change, and the cell can respond rapidly to touch, stretch, or shear. In sensory neurons, mechanosensitive channels underpin touch perception. In many other cell types, they can contribute to responses to crowding, flow, and substrate stiffness.

What makes Piezo especially interesting for mechanobiology is that its gating cannot be understood by membrane physics alone. The cortex can protect or amplify mechanosensation. If the membrane is tightly tethered to a tense cortex, the bilayer may experience different stresses than if it were free. The cytoskeleton can also reshape local curvature and tension, potentially influencing which channels open and when. In turn, the ion flux through channels can regulate the cytoskeleton by altering calcium levels and signalling pathways that control actin dynamics and myosin contractility. The *living skin* is thus a feedback circuit: mechanics influences ion flow; ion flow influences mechanics.

Cortical mechanics can also be felt on a human scale in a way that is both sobering and clinically relevant. Many cancer cells show altered cortical tension and membrane-cortex coupling compared with their healthy counterparts. A cell that is too stiff may struggle to invade; a cell that is too soft may squeeze through gaps more easily but also risks mechanical damage. In the bloodstream, circulating tumour cells must survive shear stresses and collisions; the cortex and its attachments can affect whether the cell ruptures or survives. Likewise, immune cells rely on rapid adjustments of cortical tension to crawl, to engulf pathogens, and to squeeze between tight junctions. These are not abstract mechanical properties; they influence who reaches an infection site first, who survives a metastatic journey, and how tissues maintain their barriers.

The cortex also participates in a quieter, constant task: maintaining cell shape against small perturbations. Cells in tissues are jostled by neighbours, by fluid flow, by muscle contraction, by the swelling and shrinking of extracellular matrices. The cortex, with its rapid turnover, behaves like a material that can be stiff on short timescales and fluid on longer ones. A sudden poke meets resistance; a sustained pressure leads to remodelling. This dual nature helps explain why cells can resist transient mechanical noise yet still change shape over minutes and hours during development and repair.

Even at rest, the *living skin* is rarely truly resting. Actin filaments in the cortex are continuously assembling and disassembling. Myosin motors generate pulses of tension, sometimes oscillatory, sometimes localized. Membrane-cortex linkers bind and release. Vesicles fuse and pinch off, adding and removing membrane area. The surface is alive with tiny adjustments that keep the cell near a workable mechanical set point. When conditions change—a cell enters a narrow pore, encounters a stiff matrix, or is pulled by a neighbour—the same machinery can shift the set point, altering tension, adhesion, and pressure to meet the new demand.

One can think of the plasma membrane as the cell's face to the world: where receptors listen, where signals are exchanged, where contact is made. The cortex is the facial expression: a layer of controlled tension that can stiffen, soften, wrinkle, smooth, bulge, or retract. Together they allow a cell to be simultaneously delicate and durable, to hold itself together as a pressurized bag of chemistry while remaining sensitive enough to respond to the lightest touch.

Chapter 4

# A Sense of Touch: How Cells Feel Their Surroundings

*Even without a nervous system, a single cell feels the world with exquisite sensitivity, distinguishing between the softness of brain tissue and the rigidity of bone. This tactile awareness relies on molecular machines that translate physical tension into chemical commands, allowing the cell to build a map of its environment through touch alone.*

## 4.1 The Rosetta Stone: Translating Force

A cell lives in a world that is constantly trying to deform it. Blood flow drags across the lining of arteries. Lungs stretch and relax with every breath. Skin creases, tendons tighten, a healing wound pulls at its own edges. For a long time biology treated these as mere background conditions, like the weather: important for comfort, perhaps, but not a language. The surprise of mechanobiology is that force is not just something cells endure. It is something they *read*—and then convert into chemistry.

Chemistry is the native tongue of cells. In the simplest telling, a signal is a molecule that binds to a protein receptor because its shape fits: the familiar "lock-and-key idea." Binding changes what the receptor does, which changes what other proteins do, and so on, until the cell divides, migrates, secretes, or dies. Mechanical force seems, at first glance, like the wrong kind of input. Force has no molecular "name." It is not a ligand that can be held in a pocket. It is a push, a pull, a shear, a squeeze. How can something so physical be made legible to a system that runs on molecular recognition?

The trick is that proteins are not rigid locks; they are folded objects balanced on the edge of motion. A protein is a chain of amino acids that folds into a three-dimensional shape. That folded shape is stabilized by many weak interactions—hydrogen bonds, hydrophobic packing, salt bridges—each modest on its own, together strong enough to hold a structure, but not so strong that the structure is immovable. Warmth from the environment jiggles every bond. Water wedges in and out. Proteins breathe, subtly opening and closing. In that context, mechanical force is not alien. It is simply an extra nudge on a system already poised between conformations.

Mechanotransduction is the general name for the conversion of mechanical inputs into biochemical signals. The most useful way to picture it is not as a special "force receptor" that somehow detects tension in the abstract, but as a *translation problem.* The mechanical world speaks in newtons and pascals; the cell responds in phosphorylation, ion flux, gene expression. Somewhere in between, the message must be rewritten in a molecular alphabet. The most widely used Rosetta Stone is unexpectedly simple: force changes protein shape in a way that reveals something previously hidden.

Many crucial protein features are *cryptic*—present in the sequence, but buried in the folded structure so that other proteins (or enzymes) cannot easily reach them. A binding site may be tucked into a helical bundle. A short motif that a kinase would like to phosphorylate may be pressed against the protein's core. A loop that could dock another partner may be pinned flat. Under ordinary circumstances these features are effectively invisible, like a keyhole covered by a sliding plate. The protein is not merely waiting to be bound; it is waiting to be *opened.*

Mechanical force provides a way to open it without chemistry. Pull on the ends of a folded domain and it may partially unfold. Stretch a multi-domain protein and its modules may separate like links in a chain. Press on a membrane protein and it may tilt or expand. In each case, the energy supplied by force reshapes the landscape of possible conformations and makes some states more likely than others. The most important consequence is not the dramatic image of a protein being torn apart, but the precise exposure of functional elements: a previously hidden binding motif becomes available, a phosphorylation site becomes accessible, a binding partner that could not dock now fits.

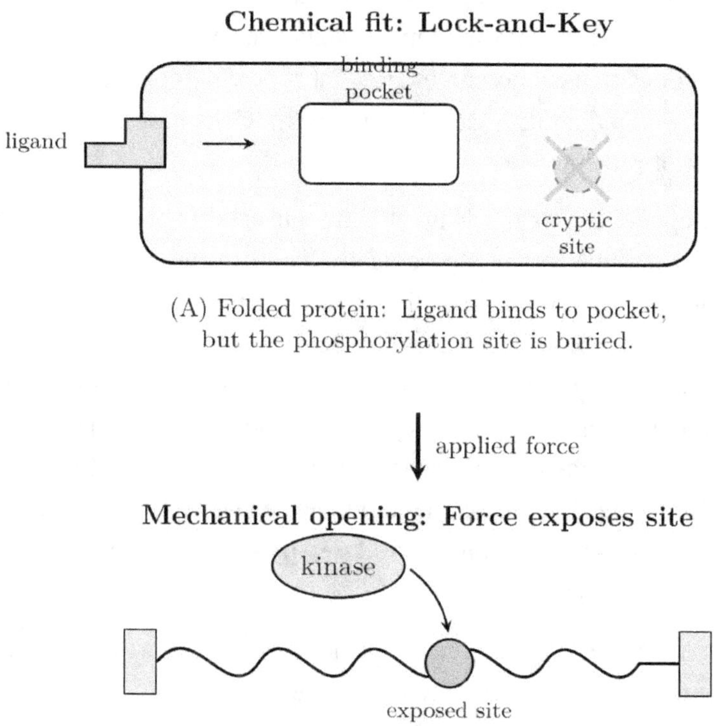


(A) Folded protein: Ligand binds to pocket, but the phosphorylation site is buried.

(B) Under tension: Unfolding reveals the site, enabling phosphorylation.

Seen this way, force is not an exotic signal. It is a means of editing access. A phosphorylation site is not created by force; it was always there. What force does is change whether the cellular machinery can *reach* it. That is why cryptic sites are such a powerful idea: they make mechanotransduction feel less like magic and more like good engineering. A protein can be designed (by evolution) to behave differently under tension than at rest, without needing a completely separate sensory apparatus.

A helpful everyday analogy is a retractable tape measure. At rest it is curled and compact; certain markings or features are buried in the coil. Pull it, and the tape straightens, exposing new numbers and allowing new measurements. The tape did not gain information when you pulled it; it gained *readability*. Cells have tape measures made of protein.

The appeal of this mechanism is its universality. In principle, any folded module that is mechanically load-bearing can be used as a translator. And cells are full of load-bearing proteins: linkers that connect membranes to internal scaffolds, scaffolds to other scaffolds, and scaffolds to the outside world. Many are built from repeating domains precisely because repeats distribute force and make extension gradual rather than catastrophic. The result is a spectrum of mechanical behaviors. Some domains behave like stiff rods. Some behave like springs. Some unfold step by step, each step unveiling a new interaction.

The language of force becomes richer when one remembers that mechanical inputs are not only about magnitude (how many piconewtons of pull), but also about timing. A tug that lasts a millisecond is different from one that lasts a minute, even if the peak force is the same. Proteins respond to *loading rate*: how quickly force is applied. They respond to *duty cycle*: how often force comes and goes. They respond to whether they are pulled steadily or jerked. These are physical features that chemistry alone struggles to represent. A cell that can translate such features into biochemical decisions gains something like a sense of touch.

A concrete example brings the abstract idea down to size. Consider talin, a large protein that acts as a physical linkage between integrins (receptors in the cell membrane that bind the extracellular matrix) and actin (a main component of the cell's internal force-generating scaffold). Talin is built in a way that makes it a likely force translator: one end can associate with the membrane-side machinery, while its long rod contains multiple helical bundles. In single-molecule experiments in which individual talin molecules were pulled with controlled force, parts of the talin rod were seen to unfold. The striking result, shown clearly by del Rio and colleagues in 2009, was that unfolding exposes binding sites for another protein, vinculin. Those vinculin-binding sites are cryptic when talin is folded. Under tension, talin becomes a better docking platform for vinculin, and vinculin then strengthens the connection to actin. A mechanical pull is rewritten as a biochemical recruitment event.

It is hard to overstate how conceptually satisfying that is. It explains how a cell can respond differently on a soft surface versus a stiff one without needing to measure stiffness as a number. If the surface yields, the force in the linkage may never reach the threshold that opens talin's cryptic sites. If the surface resists, tension builds, domains unfold, vinculin binds, and the adhesion strengthens. In that story the critical step is not a mystical stiffness sensor but a protein whose conformation depends on force.

This kind of mechanochemical switch also hints at memory. A lock that is pried open may not snap shut instantly when you let go. In proteins, the same can be true. Vinculin binding to unfolded talin can stabilize the open state, making it harder for talin to refold. That stabilizing effect creates hysteresis: the molecular state depends not only on the current force but on the recent history of force. For a cell crawling across a landscape of varying stiffness, such hysteresis is a way of smoothing a noisy world into a decision that persists long enough to matter.

Talin is only one illustration. Another, conceptually cleaner because it points directly to phosphorylation, is p130Cas, a protein involved in adhesion-associated signaling. p130Cas contains a substrate region with multiple tyrosine motifs that can be phosphorylated by Src-family kinases. Sawada and colleagues in 2006 showed that stretching this region increases its phosphorylation, consistent with the idea that extension makes the motifs more available to the enzyme. Here the translator is not a binding partner revealed, but a set of chemical targets unmasked. The same logic applies: force changes shape; changed shape changes access; access changes chemistry.

Why would evolution choose such a strategy? One reason is that it is robust. A chemical receptor must distinguish its ligand from the countless other molecules in a cell. That is a delicate recognition problem. A force translator, by contrast, can be tuned by mechanics: domain stability, linker length, the number of repeated motifs, the arrangement of helices. The input is not the identity of a molecule but the presence of tension along a particular path. That path specificity is crucial. A cell is often under many forces at once, but not all forces should trigger the same response. A pulling force along an adhesion linkage should trigger one set of behaviors; compression at the nucleus might trigger another. By placing cryptic sites in specific proteins on specific mechanical routes, a cell can make force legible without making it indiscriminate.

A second reason is speed. Chemical signaling cascades can be fast, but the earliest steps often require diffusion, binding, enzymatic turnover. Conformation changes under force can occur essentially as quickly as the force is applied, limited by molecular rearrangements that can be microseconds to milliseconds. That speed matters when a cell needs to respond to sudden shear stress in a blood vessel or a rapid touch at a nerve ending. Not all mechanotransduction is this direct—many mechanosensitive responses are slow and involve gene expression—but the initial translation can be immediate.

A third reason is efficiency. A cryptic site is an economical design. The cell does not need to synthesize a protein that is always active and then constantly suppress it. It can keep the site hidden until the mechanical condition that makes it useful arises. This is especially attractive for phosphorylation sites, because phosphorylation can be a point of no return in signaling networks. If the site is accessible only under tension, the cell avoids accidental activation when the protein is not in the right mechanical context.

There is also an emotional, almost philosophical appeal in the idea that life makes sense of its physical world by folding and unfolding. It turns the cell from a bag of reacting chemicals into something closer to a tactile machine. Yet the tactile machine is not made of gears and levers; it is made of soft matter. In the nineteenth century, long before mechanobiology had a name, physicians and anatomists noticed that tissues remodel according to how they are used: bones thicken under load, tendons strengthen with training, the heart enlarges under chronic pressure. Those observations were correct but frustratingly macroscopic. The missing link was a microscopic mechanism that could plausibly connect stress to biochemistry without hand-waving. Cryptic sites provided that link.

One historical pivot came with the rise of single-molecule manipulation techniques at the end of the twentieth century and the beginning of the twenty-first: optical tweezers, atomic force microscopy, magnetic tweezers. These tools made it possible to apply piconewton forces to single proteins and watch them respond, step by step, rather than inferring behavior from bulk averages. The point was not merely to show that proteins can unfold—biochemists already knew proteins denature—but to show that they can unfold *in a controlled, modular way* at forces that cells can generate, and that unfolding can be directly coupled to functional interactions. That was a shift from thinking of unfolding as damage to thinking of it as design.

A common misunderstanding is to imagine that mechanotransduction always requires a protein to be dragged out like taffy. In practice, the conformational changes can be subtle: a helix can rotate, a loop can lift, a domain interface can loosen. Even small changes can have large biochemical effects because enzymes are picky about geometry. A kinase does not merely need its target motif to exist; it needs it positioned and exposed long enough to catalyze a reaction. A binding partner may require a short linear motif to be presented in an extended form. Force can supply that presentation.

Another misunderstanding is to assume that the mechanical translator must be the protein that bears the most force. Often the force-bearing element and the sensing element are distributed. A stiff filament may transmit force to a more compliant domain that actually changes conformation. Or a complex may be arranged so that a small change in one element relieves steric hindrance in another. Mechanotransduction is rarely a single bolt that tightens; it is more like a tent where pulling one rope changes the tension in many others.

It also matters that forces inside cells are not imposed only from the outside. Cells pull on their surroundings using molecular motors, especially myosin motors that slide along actin filaments. That internal contractility is not mere muscle-like exertion; it is also a probing action. By pulling, a cell tests how much the environment yields. The resulting pattern of tension across its linkages determines which cryptic sites open and which stay closed. In other words, mechanotransduction is often an active sense, closer to how a hand explores an object than how a thermometer passively warms.

A final layer of elegance comes from the fact that cryptic-site mechanisms can be combined with classical chemical logic. A protein may require both force and a ligand to become active. Force might expose a site, but phosphorylation might still require a kinase that is present only under certain conditions. Or a binding site might become available under tension, but binding might occur only if the partner is activated by a separate pathway. This gives cells an AND gate: touch alone is not enough; touch plus context is. That is how a tissue can respond differently to the same mechanical stress during development, in adulthood, or in disease.

Mechanotransduction therefore begins with a deceptively modest claim: proteins are shaped objects that can be mechanically reshaped. From that claim flows a new view of cellular communication. Mechanical force is not a vague influence; it is a structured input with direction, magnitude, and time course. Cryptic sites turn that structured input into discrete biochemical events: a docking interaction permitted, a phosphorylation made possible, a signaling pathway initiated. A pull becomes a sentence the cell can read, written not in words but in conformations.

## 4.2 Anchors That Listen

A cell on a surface is not resting. Even when it looks still under a microscope, it is busy taking purchase, pulling, slipping, tightening, letting go, and trying again. This is not fidgeting; it is how a cell finds out what sort of place it has landed in. The outside world can be as soft as brain tissue, as tough as tendon, as slick as a basement membrane, as ragged as a wound bed. Chemistry tells the cell what molecules are nearby, but mechanics tells it whether those molecules are presented on something that can bear load. The structures that perform this two-way conversation are called *focal adhesions*: dense clusters of proteins that span from the extracellular matrix outside the cell, through the membrane, into the cell's internal scaffolding.

The name is a little misleading if one imagines a tidy rivet. A focal adhesion is more like a busy dock where ropes are tied, winches are running, and new cleats are being bolted down as the ship tugs against the pier. It is an anchor, but an anchor that listens. It transmits force, and the act of transmitting force changes what the anchor is made of.

The molecular hook that begins the whole arrangement is the *integrin*. Integrins are receptors embedded in the cell membrane. Their job is to bind proteins in the extracellular matrix (the meshwork of collagen, fibronectin, laminin, and many other components that give tissues their physical character). Integrins have a part outside the cell that can clasp a matrix ligand, and a part inside the cell that can connect to the internal framework. That simple geometry—outside binding and inside coupling—already hints at a mechanical function. If a cell pulls on its internal skeleton, that pull can be delivered through integrins into the matrix. If the matrix resists, tension builds across the integrin connection.

Yet an integrin by itself is not an anchor you would trust. The cytoplasmic tails of integrins are short. Their direct grip on the internal scaffolding is indirect and heavily regulated. Cells need adhesions that can be assembled quickly, reinforced when useful, and dismantled when it is time to move. Focal adhesions solve this by being modular. They are not a single protein but a shifting coalition that can number in the dozens or hundreds of distinct components. Some are mechanical linkers; others are enzymes; others are scaffolds that recruit more scaffolds. The remarkable feature is that this crowded molecular assembly remains sensitive rather than sluggish. It is sensitive because the assembly is not static. It is tuned by force.

The easiest way to picture this is to take seriously the idea of a *molecular clutch*. In a car, a clutch connects an engine to the wheels. When the clutch engages, the engine's torque is transmitted, and the wheels move. When it slips, the engine spins without moving the car. Cells have an analogous arrangement. Inside the cell, actin filaments and myosin motors generate contraction, a kind of slow, persistent tugging. If that tugging is coupled to integrins that are bound to the matrix, the cell transmits force to the outside world. If the coupling is weak or transient, the actin machinery can keep cycling without building much external traction.

This clutch metaphor does more than provide a helpful mental picture; it explains why adhesions are such capable sensors. The cell is not passively detecting stiffness in the abstract. It is actively pulling and watching what happens to the linkage. Does the connection hold long enough for tension to rise? Does it fail quickly, like trying to pull on wet sand? Does it hold, but in a way that causes parts of the linkage to open, recruit reinforcements, and become stronger? The answers depend on the physics of the environment and the chemistry of the proteins—and the focal adhesion sits precisely at that intersection.

A focal adhesion begins modestly. A few integrins bind ligands in the matrix. On the inside, an adaptor called *talin* is one of the first major connectors to arrive. Talin can bind integrin tails and also associate with actin. In the previous discussion of cryptic sites, talin served as a clear example of how force can change a protein's shape to expose new binding opportunities. That same property makes talin a plausible core of the clutch: when tension develops across talin, it can reveal docking sites for *vinculin*, another adaptor that strengthens the link to actin. A load-bearing connection becomes, under load, a better platform for further strengthening. The anchor listens by physically changing in response to what it feels.

The environment matters because it controls how quickly tension builds. Imagine a cell pulling through a talin-based linkage. If the matrix is very soft, the matrix deforms easily; the integrin-ligand connection may never experience high tension. The clutch slips in the sense that actin flow continues but does not translate into large external force. Under those conditions, talin may not be stretched enough, for long enough, to expose the binding sites that would recruit vinculin. The adhesion remains small and transient. Now imagine a stiff matrix. The same internal contraction produces higher tension across the linkage because the matrix resists deformation. Talin is more likely to be stretched, vinculin more likely to bind, and the adhesion more likely to mature into a large, stable structure. The cell has effectively performed a mechanical test without a dedicated stiffness meter. It has let physics set the outcome of a molecular competition.

What makes this particularly convincing is that force does not merely strengthen adhesions in an indiscriminate way. Force changes *composition*. Under different mechanical regimes, focal adhesions recruit different proteins, activate different enzymes, and reorganize the actin cytoskeleton in different architectures. Reviews by Benjamin Geiger and Joachim Spatz (2009) emphasized that mechanosensing is an emergent property of adhesion growth, force transmission, and signaling feedback, not a single sensor molecule. That framing matters because it avoids a common trap: searching for one magic mechanoreceptor. Focal adhesions are better understood as assemblies that can settle into different states depending on how they are loaded.

The word “loaded” deserves attention. Force in cells is not a single number. The same peak force applied slowly versus rapidly can produce different molecular outcomes. Some bonds behave like simple Velcro: the harder you pull, the sooner they fail. Biophysicists call these *slip bonds*. But some molecular interactions counterintuitively become longer-lived when pulled within a certain range. These are *catch bonds*, and they are one of the most beguiling ideas in mechanobiology because they overturn everyday intuition. If you pull on a catch bond with the right force, it holds on *better*, at least up to a point.

A concrete example appears right at the beginning of the adhesion: between an integrin and its ligand. In 2009, Fang Kong and colleagues working with Andrés J. García demonstrated catch-bond behavior for the integrin $\alpha_5\beta_1$ binding to fibronectin. Under intermediate levels of force, bond lifetimes increased rather than decreased. The effect depended on the biochemical state of the integrin and on conditions such as divalent cations, reminding us that mechanical and chemical regulation are intertwined. A bond that lasts longer under tension is not just a curiosity; it is a way for a cell to treat pulling as a *test*. If tension stabilizes the bond, then the very act of pulling helps the cell maintain its grip long enough for downstream reinforcement to occur. If tension destabilizes the bond, the cell is less likely to commit to that contact.

This begins to explain how focal adhesions can be so sensitive to the difference between surfaces that, to our eyes, look identical. A cell does not only ask, “Is there fibronectin here?” It asks, “Is there fibronectin presented in a way that can bear the forces I generate?” A sparse coating of ligand on a rigid substrate is not the same as the same ligand on a soft gel, because the mechanical feedback determines whether bonds are stabilized, whether talin unfolds, whether vinculin binds, whether actin bundles align, and whether signaling enzymes remain in place long enough to act.

If all this sounds like an engineer’s description of a load sensor, that is not accidental. Cells have had a billion years to evolve devices that are as subtle as any built in a laboratory, but assembled from floppy molecules in warm water. The engineer’s instinct is to separate structure from signal: build one component to bear load, another to detect strain, another to produce an output. Cells often blur these roles. A focal adhesion protein can be structural one moment, signaling the next, and structural again after phosphorylation changes its partners.

A good example of this blurring is the role of phosphorylation itself at adhesions. Many focal adhesion components are substrates for kinases, including Src-family kinases and focal adhesion kinase (FAK). Phosphorylation does not merely turn proteins on or off; it rewires interaction networks. A phosphorylated tyrosine can recruit proteins with SH2 domains; a phosphorylated serine can change binding preferences or stability. Force can affect phosphorylation indirectly by changing which proteins are exposed, clustered, or held together long enough for enzymes to act. In some cases, as with the stretching-sensitive phosphorylation of p130Cas described by Sawada and colleagues (2006), the substrate itself becomes more available when extended. In other cases, force changes the local density of proteins and the residence time of kinases at the adhesion, which can be just as important as catalytic rate.

The story of focal adhesions is also a story about movement. Migrating cells form new adhesions at their front edge and release adhesions at the rear. The timing must be right. Too little adhesion and the cell cannot pull itself forward; too much and it becomes stuck. Cancer cells exploit this balance when they invade tissues. Immune cells exploit it when they squeeze through blood vessel walls. Fibroblasts exploit it when they crawl into a wound and lay down new matrix. These are not abstract consequences; they show up in health and disease in ways we can see in the clinic. A scar that contracts too strongly, a vessel wall that stiffens and remodels, a tumor that becomes mechanically abnormal—all of these involve cells that have been listening to their environment and responding with altered adhesion and force generation.

A historical anecdote captures how the field began to connect traction with signaling. In the late twentieth century, researchers began to culture cells on flexible substrates and measure the tiny wrinkles or displacements produced by cellular pulling. This kind of traction force microscopy made it impossible to keep thinking of cells as passive blobs that merely adhere. Cells were clearly exerting forces, and the forces changed with substrate stiffness and ligand type. At the same time, microscopy was revealing focal adhesions as distinct structures where integrins and many cytoplasmic proteins clustered. The pieces were visible, but the mechanism—how force could reorganize molecular composition—was not yet clear. The mechanosensitive unfolding of proteins like talin and the discovery of catch bonds at integrin-ligand interfaces supplied tangible molecular events that could plausibly sit at the heart of the phenomenon. Mechanobiology became less like a set of phenomenological observations and more like a set of testable molecular hypotheses.

The most counterintuitive aspect, once one becomes comfortable with the parts list, is that focal adhesions can behave almost like *decision-making units*. They integrate multiple kinds of information: the chemistry of the ligand (collagen is not fibronectin), the spacing and density of binding sites, the stiffness and viscoelasticity of the matrix, the cell's own contractile state, and the history of loading. The same adhesion can grow, pause, slide, or disassemble depending on these inputs. It is not that the adhesion "knows" anything; it is that the kinetics of binding and unbinding, unfolding and refolding, recruitment and release, create a system with multiple stable outcomes.

One might ask why the cell invests in such a complicated apparatus rather than simply secreting glue. The answer lies in the difference between being attached and being *attached appropriately*. A nerve growth cone navigating a developing brain must attach lightly enough to keep exploring, yet firmly enough to pull itself forward. A bone-forming cell must attach strongly and persistently to build mineralized matrix. A circulating white blood cell must attach transiently to roll along a vessel wall and then arrest abruptly at the right location. These are mechanical problems, and the solutions require adhesions that can be tuned and retuned, second by second.

Even within a single cell, focal adhesions are not uniform. Some sit at the ends of thick actin bundles and bear high tension; others are small and dynamic. Their positions and orientations matter. A cell that senses an aligned fibrous matrix often becomes elongated and migrates along the fibers, a behavior that shapes how cancer spreads and how wounds heal. That guidance can be understood as a distributed outcome of many adhesions experiencing different mechanical feedback depending on their alignment with the main axes of internal contractility and external stiffness.

If focal adhesions are anchors, they are anchors tied to an active engine. Myosin motors generate contractile forces by pulling actin filaments past one another. That contractility is itself regulated by signaling pathways, including Rho-family GTPases, which respond to adhesion signaling. The result is feedback: adhesions transmit force; force changes adhesion composition and signaling; signaling changes the force-generating machinery. In a stable state, the feedback can maintain a steady level of tension that keeps an adhesion mature. In a changing environment, the feedback can drive rapid remodeling. This is how a cell can switch from a rounded, loosely adherent state to a spread, strongly adherent one when it encounters a stiff substrate.

A final detail makes the "listening" metaphor feel literal: focal adhesions often produce signals only when there is tension to keep the signaling machinery assembled. Many of the relevant proteins interact through low-affinity binding that becomes significant only when molecules are clustered. Force can stabilize clustering by favoring conformations that recruit additional partners, and by keeping the adhesion from disassembling. The adhesion becomes a physical platform that exists only as long as mechanical conditions support it, and that platform is exactly what allows enzymes and adaptors to remain in proximity long enough to pass information along. In that sense, force does not merely trigger a signal; it sustains the stage on which signaling can occur.

When a cell grips the matrix, it is not simply holding on. It is asking a question with its own muscles and reading the answer in the shifting architecture of proteins at the membrane. The astonishing part is that the answer is not delivered as a number but as a change in who can bind whom, who gets phosphorylated, who stays, who leaves. An anchor that listens does so by becoming, under tension, a different anchor.

## 4.3 The Fastest Reflex

Press a fingertip against a table and the sensation feels immediate. There is no perceptible delay while the body runs chemical computations. Something opens, electricity moves, and the nervous system is informed in real time. That speed is not a luxury. Without it, walking would be clumsy, gripping a cup would be guesswork, and blood pressure would surge and crash with every heartbeat. A cell often needs a response that is faster than protein recruitment and slower feedback loops can deliver. The fastest way to translate force into action is to let force open a door.

That door is an ion channel. Ion channels are proteins embedded in cell membranes that form pores, allowing charged atoms—ions—to cross. Cells spend a great deal of energy maintaining different ion concentrations inside and outside, so opening a channel is like opening a sluice gate in a dam: the flow is swift and consequential. Because ions carry charge, the movement is electrical. A tiny structural change in a channel can produce a measurable current in a fraction of a millisecond. The cell does not need to synthesize a second messenger or build a multiprotein scaffold. The signal is built into physics.

Mechanically activated ion channels provide this ultrafast route. They open when a membrane is stretched, curved, compressed, or otherwise stressed. For decades, physiologists could record such currents in cells and infer that some molecular gate must be responding to force, but the identity of the gate remained mysterious. It is one thing to observe that a membrane patch produces an electrical response to poking; it is another to name the protein that does the opening.

The breakthrough came in 2010 when Ardem Patapoutian's group, led in the key experiments by Bertrand Coste and colleagues, identified a large protein required for mechanically activated currents. They named it Piezo, from the Greek for pressure. Soon it became clear there were two main family members in mammals: Piezo1 and Piezo2. With a name and a sequence, the field could finally connect molecular structure, cellular function, and whole-organism physiology. The sense of touch, the regulation of blood flow, the feeling of lung inflation, the detection of bladder fullness—all could now be traced, in many contexts, to the opening and closing of a specific kind of pore.

Piezo channels are not subtle little holes. They are among the largest ion channels known, enormous assemblies of protein that sit in the membrane like elaborate machines. Their scale hints at their trick. A small protein might respond to force if it had a clever hinge, but a large protein can respond to force simply by being a major architectural feature of the membrane. Piezo does not just occupy the membrane; it reshapes it.

To appreciate why that matters, it helps to think about the membrane itself as an elastic sheet. A cell membrane is a bilayer of lipids: two layers of fat-like molecules with watery heads on the outside and greasy tails tucked inside. It is thin, flexible, and under tension. Stretching the membrane changes the energetic cost of making space for a protein conformation. Some channel states occupy a larger area in the membrane than others. When tension is high, a conformation that spreads the membrane less can be penalized, while a conformation that relieves local curvature or accommodates stretch can be favored. In that way, the membrane is not merely a backdrop; it is part of the gating mechanism.

This leads to a central debate about mechanogating, often summarized as *force-from-lipids* versus *force-from-filaments*. In the force-from-lipids picture, the membrane's own tension and curvature act directly on the channel. No special tethers are required. The channel is sensitive because it is embedded in a deformable medium. In the force-from-filaments picture, forces are transmitted through attachments to the cytoskeleton inside the cell or to the extracellular matrix outside it, so that pulling on the cell's internal cables can tug the gate open. These are not mutually exclusive. A channel may be primarily driven by membrane tension in one context and by tethering in another, and different channels may sit at different points along this spectrum. What Piezo has done is provide a concrete test case: a mechanosensitive channel with a distinctive architecture that invites membrane-centered explanations, yet lives in cells where cytoskeletal coupling can never be completely ignored.

The structure of Piezo reinforces the idea that it is a membrane sculptor. High-resolution studies over the past decade have revealed Piezo as a trimer: three similar subunits arranged around a central pore. The overall shape has been compared to a three-bladed propeller, or a triskelion, and the key functional detail is that the complex imposes curvature on the membrane. Rather than sitting in a flat bilayer like a coin in a sheet, Piezo bends the sheet. The membrane around the channel is pulled into a dome-like deformation. That deformation stores elastic energy. Under increased tension, the membrane resists being bent; flattening becomes favorable. The channel, by switching between curved and flatter states, can couple membrane tension to pore opening.

One can picture it with a simple tactile analogy. Imagine pressing a suction cup onto a window. The rubber dome stores energy as it is deformed. If you pull on the rim, the dome flattens and the forces redistribute. The suction cup's shape change is not merely passive; it changes what the cup can do. Piezo is not a suction cup, but it shares the principle that curvature can be converted into a mechanical trigger.

When Piezo opens, it allows cations to enter the cell, including calcium ions ($Ca^{2+}$). Calcium is especially potent because it serves as a universal intracellular signal. Cells keep cytosolic calcium low, so even a small influx can act like a flare, setting off downstream events. Calcium can activate enzymes, alter cytoskeletal dynamics, and influence gene expression. This provides a bridge from the fastest reflex to slower adaptations. An ion channel can open in milliseconds, but the consequences can play out over minutes or hours.

The timescale difference matters because it clarifies why cells use different mechanotransduction strategies for different problems. Focal adhesions are superb for integrating sustained forces, measuring stiffness through repeated tugging, and reorganizing the cytoskeleton in a context-dependent way. They are deliberative, in a mechanical sense. Piezo channels are suited to sudden changes: a quick indentation, a pressure pulse, a stretch that arrives too fast for an adhesion to remodel. In the body, many of the most important mechanical events are pulses. A heartbeat is a pulse. A footstep is a pulse. A cough is a pulse. A sensory system that cannot keep up with pulses is not much of a sensory system.

A concrete example is blood flow in arteries. As each heartbeat sends a pressure wave through the vessel wall, endothelial cells (the cells lining blood vessels) experience shear stress from flowing blood and stretch from the pulsatile expansion of the vessel. These mechanical cues are not incidental. Endothelial cells must respond to them to maintain vessel tone, regulate inflammation, and remodel the vessel wall. Piezo1 has been implicated as a key mechanosensor in endothelial cells, contributing to calcium influx and downstream signaling that helps the vasculature adapt to flow. When this mechanosensitivity is disturbed, the consequences can be systemic, influencing blood pressure regulation and vascular development.

Another vivid example comes from touch itself. Piezo2 is strongly associated with mechanosensation in sensory neurons and specialized skin cells. People with certain rare mutations affecting Piezo2 can have altered touch and proprioception (the sense of body position). Proprioception is a particularly striking case because it is so easy to overlook until it is compromised. Closing your eyes and touching your nose feels effortless, but it depends on a constant stream of mechanical information from muscles and joints. That stream must be fast and continuous. The body cannot afford a slow chemical deliberation each time a muscle spindle is stretched. Ion channels, opening and closing with mechanical deformation, provide a direct line from force to nerve firing.

The fascination of Piezo is also that it makes mechanotransduction feel less like a metaphor. With focal adhesions, one can be tempted to treat "sensing" as a poetic description of complex molecular assembly. With Piezo, sensing is literally a gate opening under force. One can measure the current. One can see the channel's state change. A physical perturbation becomes an electrical event.

Yet even here, biology resists being reduced to a single switch. Mechanical stimuli come in many forms: indentation, stretch, shear, osmotic swelling that increases membrane tension from within. Piezo channels must discriminate among these, and cells must decide what to do with the resulting calcium signal. A calcium spike can trigger secretion in one cell type, contraction in another, gene expression in a third. The channel is a translator, but not the whole sentence.

The membrane itself adds nuance. Lipid composition varies between tissues and even between regions of a single cell membrane. Cholesterol content, the length and saturation of lipid tails, and the presence of specialized lipid domains can all affect membrane stiffness and curvature preferences. Because Piezo gating depends on the membrane's mechanical properties, changing the membrane can change sensitivity. This means that mechanosensation is not only about the channel protein but also about the material it sits in. A neuron's membrane is not mechanically identical to an endothelial cell's membrane, and a healthy membrane is not identical to one altered by inflammation or metabolic disease. The same channel can behave differently in different physical contexts.

Cytoskeletal coupling adds another layer. Even if a channel is primarily gated by membrane tension, the cytoskeleton can influence local tension by corralling the membrane, pulling on it, or buffering it against deformation. Conversely, calcium influx through Piezo can rapidly remodel the cytoskeleton, altering cell stiffness and changing how future forces are transmitted. This makes the force-from-lipids and force-from-filaments pictures less like competing ideologies and more like different slices through a feedback system.

It is also worth noticing how the discovery of Piezo changed the emotional texture of the field. Before a molecule is named, a phenomenon can feel like a ghost: real in measurements, elusive in mechanism. Once the gene is identified, experiments proliferate. Mutations can be linked to symptoms. Tissue-specific knockouts can reveal developmental roles. Structural biology can reveal plausible gating mechanisms. The discovery is not merely a technical step; it turns an idea into a handle. Patapoutian's identification of Piezo (2010) gave mechanosensitive currents a molecular face, and with that came a surge of clarity about how deeply mechanosensation is woven into physiology.

The physiological reach is broad, but the conceptual reach may be broader. Piezo channels demonstrate a principle: mechanotransduction can be *direct.* Force can act on a protein in a way that produces an immediate change in permeability, which produces an immediate electrical and chemical signal. This is mechanotransduction with minimal intermediates. In a world where much of cell biology is taught as cascades of binding partners and phosphorylation events, it is refreshing to encounter a mechanism where the first step is simply geometry.

That simplicity does not make the system crude. Piezo channels can adapt. Many mechanosensitive currents show inactivation: the channel opens under a stimulus and then closes even if the stimulus remains. This is the cellular equivalent of sensory adaptation, preventing saturation and preserving sensitivity to changes rather than absolute levels. The details vary by channel and context, but the point is general: even the fastest reflex is not merely a binary on/off response. It is shaped by kinetics, by membrane mechanics, and by the cell's internal state.

A final image helps tie the fast reflex back to the slower forms of mechanotransduction without collapsing them into one. Focal adhesions are like hands: they grip, test, and reorganize. Piezo channels are like whiskers: they flick and report, instantly, when something brushes by. A hand can learn the weight of an object by lifting it; a whisker can detect the lightest touch before the hand has even moved. Cells need both. A vessel-lining cell needs immediate awareness of a pressure pulse and longer-term remodeling of its attachments. A sensory neuron needs rapid gating to fire action potentials and longer-term changes in excitability and gene expression when mechanical environments change. The same organism that relies on instantaneous touch also relies on the slow sculpting of tissues under chronic load.

A membrane stretched by a heartbeat, a skin cell indented by a fingertip, a bladder wall distended by filling: in each case, tension spreads across a bilayer only a few nanometres thick. Piezo turns that tension into a pulse of ions, and a pulse of ions into information, fast enough to keep a body coordinated with the physics it inhabits.

Chapter 5

# Fate by Force: Stiffness and Cellular Decision-Making

*Cells do not operate in a vacuum; they function on substrates ranging from the squishy softness of brain tissue to the unyielding rigidity of calcified bone. This chapter reveals how cells actively probe this resistance, translating physical feedback into the biological destiny of becoming nerve, muscle, or bone.*

## 5.1 Soft Beds and Hard Floors

A human body is a map of mechanical landscapes. Brain tissue yields under a fingertip; cartilage pushes back; bone barely notices. That range is not a side effect of biology. It is part of the information biology uses.

One of the most unsettling ideas in modern cell biology is that a stem cell can be steered without changing its genes, without adding a growth factor, sometimes without even changing the chemistry of what it touches. Change the *feel* of the surface beneath it—how much it gives way when the cell pulls—and the cell can begin to build the machinery of a neuron, a muscle fibre, or an osteoblast (a bone-forming cell). The instruction is not a molecular keyword but a mechanical one: *stiffness*.

Stiffness is a deceptively everyday word. In mechanics, the stiffness of a material describes how much it deforms under a given force. A useful number is the elastic modulus, often measured in kilopascals (kPa) for soft biological tissues. Roughly speaking, brain-like softness sits around fractions of a kPa to a few kPa; muscle is in the tens of kPa; pre-calcified bone matrix is stiffer still. Cells do not carry pocket calculators, but they do pull, and they can register how far the world moves when they do.

The turning point came when researchers learned how to present cells with surfaces that were chemically similar but mechanically different. The trick was not exotic. Polyacrylamide gels—close cousins of the materials used in electrophoresis, familiar to many laboratories—can be made soft or stiff by changing the amount of crosslinking. They can then be coated with the same adhesive proteins (such as collagen) so that, to the cell, the chemical footholds look comparable. What changes is the compliance: on a soft gel, the cell's pull sinks in like a hand pressing into a mattress; on a stiff gel, the same pull is met with resistance, like pushing on a tabletop.

In 2006, Adam Engler and colleagues made the now-classic observation that mesenchymal stem cells, when cultured on such elasticity-tuned substrates, tended to adopt fates that matched the stiffness of the material beneath them. On very soft, brain-like gels, they expressed markers associated with neural lineages. On intermediate, muscle-like stiffness, they leaned toward myogenic programmes. On stiffer, bone-like substrates, they upregulated osteogenic markers. The provocative part was not that chemical context matters—everyone expected that. The provocation was that *mechanical context alone* could bias the decision.

The result feels almost too neat: soft makes nerve, medium makes muscle, hard makes bone. Biology is never that clean. Not every cell obeys. The process takes time. And, crucially, there is a window of plasticity early on when soluble signals can redirect the trajectory. Yet the experiment forced a new question into the centre of developmental thinking: if cells can read stiffness, how do they do it? What is a "stiffness sensor" made of?

A cell has no rigid skeleton like a vertebrate's bones. Its mechanical frame is a dynamic meshwork of protein filaments. The most important for this story are actin filaments, which can be pulled by myosin motors, generating contractile forces much like miniature muscle. When a cell attaches to a surface, it does so through integrins: transmembrane proteins that bind extracellular matrix outside and connect, through adaptor proteins, to actin inside. These attachment sites, called focal adhesions, are not nails hammered into a wall; they are living assemblies that grow, shrink, and change composition depending on load.

Now imagine the cell as an impatient tester. It reaches out, grabs the matrix through integrins, and tugs. On a stiff substrate, the tug does not move the world much. The resistance allows tension to build in the actin network. On a soft substrate, the same tug mostly deforms the gel; tension inside the cell does not build as effectively. The distinction is not just the displacement outside the cell; it is the *level of internal tension* the cell can sustain.

Internal tension matters because many proteins in adhesions and the cytoskeleton change their behaviour under force. Some unfold to expose binding sites. Some bind more strongly when pulled (so-called catch bonds). Some recruit additional proteins only when stretched. These are not metaphorical switches: they are molecules whose shapes, and therefore interactions, are altered by mechanical load. Over minutes to hours, a stiff environment tends to promote larger, more stable adhesions and thicker actin bundles (stress fibres). A soft environment tends to favour smaller adhesions and a more delicate, cortical actin arrangement.

Engler's group added a key piece by showing that inhibiting non-muscle myosin II—reducing the cell's ability to generate contractile tension—blocked stiffness-directed lineage specification. In other words, the cell's reading of stiffness requires the cell to *pull*. The "sensor" is not a passive strain gauge embedded in the membrane. It is an active test: force generation and force sensing are coupled.

That coupling makes intuitive sense if you have ever tried to judge the ripeness of fruit. You do not learn much by looking. You squeeze. The cell's squeeze is slower and smaller, but the logic is the same: apply a force and measure the response. The measurement is distributed across adhesions, actin, and the membrane. The readout is not a number; it is a pattern of molecular states that either stabilise or dissolve, amplify or dampen.

An ordinary clinical scene can make the stakes feel real. Consider what happens after a deep cut. Blood clots, then cells migrate in to rebuild tissue. The environment they enter is not mechanically neutral. A fibrin clot is soft and provisional. As repair proceeds, fibroblasts deposit collagen, and the matrix stiffens. If stiffness can bias fate and behaviour, then the mechanical evolution of a wound is not just scaffolding; it is instruction. Too much collagen and crosslinking, and the stiffening can overshoot into fibrosis: scar tissue that is mechanically and functionally different from what it replaces. Even without naming specific diseases, the principle is clear: when the "bed" beneath cells hardens, their choices can change.

One of the earliest hints that cells take mechanical gradients seriously came even before the stem-cell lineage work. Around 2000, Chien-Min Lo and colleagues showed that cells can migrate preferentially toward stiffer regions of a substrate, a behaviour later popularised as durotaxis. That discovery reads today like a quiet preface: if a cell can crawl up a stiffness gradient, then it must be able to compare "here" and "there" mechanically, even in the absence of a chemical trail. In living tissues, chemical and mechanical gradients often coincide, but they do not have to. A stiffening matrix can become a guidance cue.

The mesenchymal stem cell experiments sharpened the question into something more personal: how can the same genome produce such different cell types, and how can the mechanical environment tip the balance?

To see why stiffness can feel like fate, it helps to picture a cell not as a bag of chemicals but as a pre-stressed structure. Engineers use the word *tensegrity* for systems whose stability comes from a balance of tension and compression. Cells are not perfect tensegrity models, but the metaphor captures something important: the cytoskeleton is under tension even at rest, and changes in that tension reshape the cell. Shape is not cosmetic. It controls which proteins assemble where, how the nucleus is deformed, how membranes are stretched, and which signalling pathways remain active long enough to change gene expression.

On a soft substrate, stem cells tend to spread less. They look rounder, with fewer prominent stress fibres. The nucleus, too, is often less flattened. On a stiffer substrate, the same cells spread more extensively, build stronger stress fibres, and adopt a more polarised, taut appearance. Over days, those differences in physical state can become differences in transcription: which genes are turned on, which are silenced, which epigenetic marks are laid down. Biochemistry is still the language, but mechanics can choose which sentences are easy to write.

There is also an important subtlety: stiffness is not a single property. Many biological materials are viscoelastic, meaning they behave partly like elastic solids and partly like viscous fluids. A gel can be "soft" in the sense that it deforms easily at first, but "stiff" in the sense that it resists sustained deformation, or vice versa. Tissues can also relax stress over time. Cells pull repeatedly, not once. So a cell's experience of stiffness depends on timescale: the same matrix may feel different under a rapid tug than under a slow, sustained contraction. Early experiments often treated stiffness as the main variable, and that simplification was fruitful, but real tissues add these temporal textures.

Still, the core of the story remains: stem cells do not merely occupy a mechanical environment; they interrogate it. The interrogation has consequences for lineage.

A concrete way to understand the lineage bias is to think about what different cell types must be *good at* mechanically. Neurons extend long, delicate processes and function in an exceptionally soft environment; excessive tension and rigidity are not their native conditions. Muscle cells must build aligned contractile machinery and transmit force; an intermediate stiffness encourages the formation of aligned actin structures and robust adhesions appropriate to that role. Osteoblasts ultimately help produce and remodel a stiff mineralised matrix; commitment to that fate on a stiff substrate seems almost like practice: the cell learns to live in a world that resists it.

This "matching" interpretation can be taken too literally. Cells are not consciously training. Yet selection and development can favour circuits in which mechanical context stabilises the programmes that best fit that context. A stem cell that begins to express bone-related genes on a stiff substrate may also increase its contractility and adhesion strength, reinforcing the very tension that keeps those genes active. That is how a bias becomes a commitment.

The historical anecdote that often gets overlooked is how much of this depended on materials science and a willingness to treat the cell's culture dish as a controllable physical environment rather than a neutral stage. For decades, standard cell culture plastic or glass provided an extremely stiff surface, orders of magnitude stiffer than most tissues. Many "normal" behaviours observed in vitro were, in hindsight, behaviours on something closer to a rock than to any organ. The elasticity-tunable gel was not merely a new technique; it was a corrective lens. It revealed that what looked like intrinsic cellular preference could be an artefact of the floor.

That insight carries a quiet ethical resonance. If the environment can steer fate, then the culture conditions used in regenerative medicine are not just supportive; they are directive. When researchers attempt to grow replacements for cartilage, cardiac patches, or bone grafts, the mechanical character of the scaffold is not a passive container. It is part of the recipe. A scaffold that is too stiff, too soft, or that relaxes stress too quickly can lead cells to build the wrong tissue, or to build the right tissue poorly.

Mechanical instruction also complicates the dream of a universal stem cell that can be "told" what to become with a cocktail of growth factors. Chemical cues matter profoundly, but mechanical cues can veto or distort them. In Engler's work, there was an early period when soluble factors could redirect outcomes before a later commitment. That temporal feature is worth holding onto. Fate is not a single switch thrown at a single moment. It is a process with inertia. Mechanics can load that process in one direction, making it harder to reverse later.

A final piece of intuition helps: cells do not sense stiffness at a distance. They sense it through contact, through adhesion, through the immediate deformation they cause. That makes stiffness a local, tactile signal. Two neighbouring cells in the same tissue can experience different mechanics if one is attached to a collagen fibre bundle and the other to a softer region. A tumour, for example, can create stiff tracks of aligned collagen that guide migration while adjacent regions remain softer. Even without invoking any particular pathology, the general lesson is that the mechanical microenvironment is patchy, and cells read it with the granularity of their own attachments.

The deep surprise, then, is not only that stem cells can be guided by softness or hardness, but that such guidance can be achieved with a single physical parameter while leaving chemistry largely constant. That is what makes stiffness feel like a language: a code written in resistance and compliance, spoken through integrins and actin, and heard in the nucleus as altered gene expression.

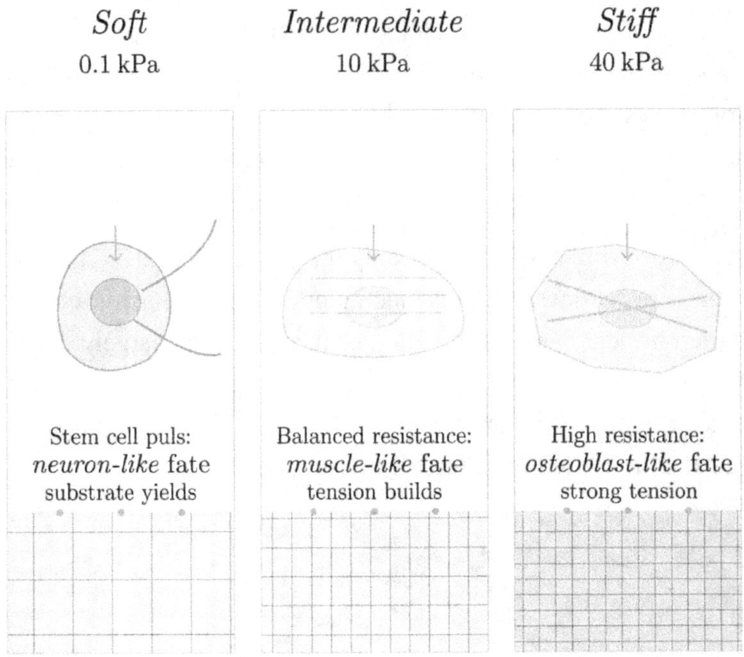


Same adhesive coating; different elasticity biases lineage.

The diagram is a cartoon, but it encodes a serious claim: the cell's internal machinery becomes different because the cell's *physical conversation* with its substrate becomes different. In a soft environment the cell's pulls are swallowed by deformation, and the cell lives in a lower-tension state. In a stiff environment the cell's pulls are reflected back, building tension and stabilising a different architecture. Those are not merely mechanical phenotypes; they are gateways into different gene-expression regimes.

If all this sounds like a modern idea, it is worth remembering that surgeons and physiotherapists have long worked with an intuition that tissues respond to load. Bone thickens where it is stressed; muscles waste when they are not used. Mechanobiology brings that intuition down to the scale of a single adherent cell making a decision. The stem cell on a gel is not performing exercise, yet it is still living in a world where force and form are linked. When the floor is soft, the cell behaves as if it belongs in softness. When the floor is hard, the cell behaves as if it must become something that can make a life on hard ground.

## 5.2 The Messenger in the Nucleus

A cell can pull on its surroundings all day long and still fail to change its life unless the tug is converted into something the genome can understand. Actin fibres and focal adhesions are excellent at feeling and generating force, but they are not where lasting decisions are written. Lasting decisions are written in transcription: which genes are copied into RNA, which programmes are amplified, which are kept silent. The remarkable trick of mechanobiology is that a physical condition outside the cell can, hours later, show up as a different pattern of gene expression inside the nucleus.

One way to imagine the problem is to picture two rooms separated by a guarded door. Outside is the cytoplasm: busy, noisy, full of scaffolding and motors. Inside is the nucleus: guarded, selective, and crowded with DNA. If mechanical tension is going to influence transcription, something must cross that door or change how the door behaves. For many years, mechanotransduction was discussed as if it were mainly a chemical relay: force activates a kinase at an adhesion, kinases phosphorylate other proteins, a transcription factor is modified, and eventually a gene turns on. That story is real, but it is incomplete. In some cases, the "messenger" is less like a chemically altered signal and more like a traveller whose location is controlled by tension.

YAP and TAZ are two such travellers. Their names are shorthand for "Yes-associated protein" and "transcriptional co-activator with PDZ-binding motif," which is not dinner-party language. The important point is that neither is a typical transcription factor that binds DNA on its own. YAP and TAZ are co-activators: they enter the nucleus and partner with DNA-binding proteins (notably TEAD family transcription factors) to turn on sets of genes involved in growth, survival, and the construction of a force-bearing cell. When they are kept out of the nucleus, those gene programmes are muted. When they enter, they can help rewrite a cell's priorities.

What makes YAP/TAZ special in the mechanobiology story is how strongly their location depends on mechanics. In 2011, Stefano Piccolo's group, led by Francesca Dupont, gave the field a clear demonstration: when cells are spread on stiff substrates, YAP/TAZ accumulate in the nucleus and become active; when cells are confined, poorly spread, or placed on soft matrices, YAP/TAZ are excluded from the nucleus and their activity drops. The switch was not subtle. Under the microscope, it could look like a dramatic change in address: a protein that is mostly nuclear in one condition becomes mostly cytoplasmic in another.

That observation carried a whiff of inevitability. If stiffness can bias stem-cell fate, then a system that turns mechanical tension into a nuclear programme for proliferation and survival is exactly what one would expect evolution to maintain. A stiff environment is often a sign of dense tissue, strong matrix, and the need to generate traction. It can also be a sign of damage, scarring, or tumour-associated remodelling. In such settings, a cell that interprets stiffness as "build more structure, resist stress, keep growing" may gain an advantage. In a soft environment like brain tissue, the same programme might be hazardous, prompting unchecked growth or inappropriate stiffening.

So the cell keeps a gatekeeper, and tension is one of the keys.

The gate is not a single molecular latch. YAP/TAZ sit at the intersection of several inputs: cell shape, the architecture of the actin cytoskeleton, the activity of Rho-family GTPases that regulate contractility, and the Hippo pathway, a kinase cascade that can mark YAP/TAZ for retention in the cytoplasm and degradation. For a popular account, the most useful simplification is this: *high tension tends to favour YAP/TAZ in the nucleus; low tension tends to keep them out.* "Tension" here is not an abstract feeling. It is the contractile state of actomyosin networks pulling against adhesions and the substrate.

The reliance on tension leads to a pleasantly concrete picture. A cell on a stiff matrix spreads widely. Its actin fibres become organised into thick bundles, and myosin motors pull along them. The focal adhesions at the ends of those fibres mature and strengthen. That architecture is mechanically capable of transmitting force toward the cell centre, and ultimately to the nucleus, which is not a floating orb but a physical object connected to the cytoskeleton. As tension rises, YAP/TAZ are more likely to enter the nucleus and stay there. Once inside, they promote the transcription of genes that support cytoskeletal assembly, extracellular matrix interactions, and survival under mechanical stress. The cell becomes better at doing the very things that maintained the tension in the first place.

A historical anecdote helps locate why this was such a conceptual shift. Before YAP/TAZ were widely recognised as mechanosensitive regulators, many biologists were already familiar with a different mechanotransduction axis that depended on actin dynamics: SRF (serum response factor) and its co-activator MRTF-A. SRF controls genes involved in cytoskeletal proteins and immediate early responses. MRTF-A is held in the cytoplasm by binding to G-actin (monomeric actin). When actin polymerises into filaments, the pool of free G-actin drops, MRTF-A is released, and it can enter the nucleus to assist SRF-driven transcription. In that older story, force and actin remodelling can indirectly influence transcription by changing the balance of actin monomers and filaments. It was a clever mechanism, but it could still feel like a chain of biochemical consequences of mechanics rather than a direct mechanical gate.

YAP/TAZ brought a different flavour: the location of a co-activator seemed to respond crisply to whether a cell was "tense and spread" or "relaxed and round." Researchers could perturb the system with pharmacological inhibitors of myosin, with micro-patterned islands that restricted cell spreading, or with gels of different stiffness, and YAP would move. The nucleus began to look less like an insulated library and more like a room whose doors and walls shift under load.

That is not merely a metaphor. In 2017, Pere Roca-Cusachs and colleagues, including Arnau Elosegui-Artola, provided evidence for a strikingly physical route into the nucleus. They argued that when cells spread on stiff substrates, actomyosin tension can flatten the nucleus. That deformation, in turn, can stretch nuclear pores—the massive protein complexes that control traffic in and out of the nucleus—and alter transport in ways that favour YAP nuclear entry. The implication was bold: the nuclear pore itself might behave as a mechanosensor, changing its permeability or transport characteristics when the nucleus is mechanically deformed.

It is easy to misunderstand this claim if one imagines nuclear pores as simple holes. They are not holes; they are elaborate machines, built from many proteins (nucleoporins), with selective barriers and active transport. But like many protein machines, they have mechanical properties. If you deform the scaffold they sit in, you may change how easily cargo moves through. The idea that force might act at the level of nucleocytoplasmic transport is one of those moments when biology suddenly sounds like engineering, and it also offers an intuitive explanation for why stiffness has such a dramatic effect. A flattened nucleus is not merely a readout of tension; it can become a control knob on which messengers are allowed in.

How does the cell pull on the nucleus in the first place? It helps to introduce another piece of cellular machinery: the LINC complex, short for "linker of nucleoskeleton and cytoskeleton." LINC complexes span the nuclear envelope. On the cytoplasmic side, they connect to actin and other filaments; on the nuclear side, they connect to the nuclear lamina and chromatin-associated structures. One can think of them as physical rivets that allow forces generated at adhesions to be transmitted to the nucleus. In 2013, studies on perinuclear actin structures, sometimes described as an actin cap, strengthened the idea that cells can route forces efficiently to the nuclear surface, enabling rapid mechanical coupling rather than slow diffusion-based signalling.

The nuclear lamina, composed largely of lamin proteins, then becomes part of the story. Lamins form a fibrous network underlying the inner nuclear membrane, contributing to nuclear shape and stiffness. Janmey, Discher, and colleagues have long emphasised that different tissues have different mechanical demands, and in 2014 work highlighted that lamin-A levels scale with tissue stiffness across tissue types. The interpretation is conceptually rich: lamin-A is both a readout of mechanical environment (stiffer tissues correlate with higher lamin-A) and a regulator, because changing lamin-A changes nuclear mechanics and can influence differentiation. If the nucleus is mechanically integrated into mechanotransduction, then its own structural tuning becomes a variable in fate decisions.

At this point, it is tempting to insist on a single "main pathway": force goes from adhesions to actin to LINC to lamina to nuclear pores to YAP. That would be too tidy. Cells almost certainly use multiple interlocking routes. YAP/TAZ are regulated by phosphorylation states that affect their binding to cytoplasmic anchors and their degradation. Actin organisation affects upstream regulators such as RhoA signalling and can modulate Hippo pathway activity. Nuclear mechanics can alter transport. These are not mutually exclusive. In a living cell, they are layered: biochemical switches and mechanical gates can act together, sometimes redundantly, sometimes synergistically.

The payoff of this complexity becomes clear when one asks what YAP/TAZ do once they are nuclear. They promote gene programmes that help cells cope with mechanically demanding environments: genes involved in cytoskeletal components, focal adhesion proteins, extracellular matrix interactions, and proliferative signals. In practical terms, a cell on a stiff substrate becomes better at spreading and pulling; a cell with nuclear YAP becomes better at maintaining the architecture that supports nuclear YAP. That circularity can feel dangerous, because it resembles a positive feedback loop. In normal development and repair, such loops can be useful: they stabilise a decision. Once a tissue region is meant to stiffen and mature, the cells in it can lock into the appropriate programme. In disease contexts, the same stabilisation can become a trap: stiffening promotes nuclear YAP, nuclear YAP promotes programmes that further stiffen or reinforce traction, and the tissue can drift toward fibrosis-like mechanics.

A concrete example brings the abstract circuit into view. Consider fibroblasts, the matrix-building cells that are recruited during wound repair. In a soft provisional matrix, they may remain relatively quiescent. As collagen accumulates and the matrix stiffens, fibroblasts can adopt a more contractile phenotype (often termed myofibroblast-like), generating stronger forces and producing more matrix. YAP/-TAZ activity is frequently implicated in this kind of transition. The details differ across tissues, but the logic is consistent: a mechanical change in the environment promotes a nuclear programme that amplifies force generation and matrix remodelling. The cell behaves as if stiffness is a command to become a builder and a puller.

The appeal of YAP/TAZ as "tension messengers" lies in their legibility. If you stain a cell and look, you can often see where YAP is. That visibility has made it a favourite readout in mechanobiology labs. But legibility is not causality. YAP nuclear localisation correlates with stiffness and spreading, but correlation alone does not prove that YAP is the decision-maker rather than a passenger. Over the past decade, experiments that manipulate YAP/TAZ levels and activity have shown that they are not merely markers: they can be required for certain mechanosensitive gene programmes, and they can push cell behaviour in ways that resemble a "stiffer" phenotype even when the environment is not maximally stiff.

The deeper point is psychological as much as technical. For a long time, the nucleus was treated as a protected vault whose contents are controlled chiefly by chemical signals, with mechanics acting at the edges. YAP/TAZ and nuclear transport mechanics encouraged a different intuition: the nucleus is a mechanically integrated organelle. When the cell pulls on the world, part of that pull is routed inward, shaping nuclear geometry, altering transport, and changing which transcriptional regulators can do their work.

This changes how one interprets cell shape. A spread cell is not simply a cell that has more area. Spreading is a mechanical state with consequences: it changes the organisation of actin, the load on adhesions, the tension transmitted to the nuclear surface, and the ease with which certain proteins can enter the nucleus. Confinement, conversely, is not merely a lack of room. It is a shift in tension that can keep growth-promoting programmes quiet. That is one reason why tiny engineered islands that restrict spreading can push cells toward differentiation or quiescence even when the chemical environment is permissive for growth.

It also changes how one interprets stiffness itself. Stiffness is not only "hardness." It is the capacity of the environment to serve as a reaction partner for cellular forces. A cell cannot build high internal tension if the outside world yields too easily. YAP/TAZ respond less to the mere existence of adhesions than to the productive resistance those adhesions encounter.

A useful way to think about YAP/TAZ is as a cellular form of common sense. If you are standing on ice, you do not sprint. If you are standing on firm ground, you may. Cells face analogous constraints: on a soft matrix, heavy contractility and large adhesions may be wasteful or destabilising; on a stiff matrix, they become workable and sometimes necessary. YAP/TAZ help translate that constraint into gene expression so that short-term mechanics become long-term phenotype.

When this translation fails or is hijacked, the consequences can be profound. A tumour is not just a genetic event; it is often a mechanical event in the surrounding tissue. Many solid tumours become stiffer than the tissue they arise from, in part through altered extracellular matrix deposition and crosslinking. In such a stiffened niche, nuclear YAP/TAZ activity can support survival and proliferation. That does not mean YAP causes cancer in a simplistic way, but it illustrates why mechanics can become entangled with growth control. The question is no longer only "what mutations occurred?" but also "what mechanical neighbourhood did those mutations grow up in?"

The nucleus, then, is not the end of the mechanobiology story; it is the point where force becomes interpretation. YAP and TAZ are not the only interpreters, but they are among the most persuasive because they behave like couriers whose route is opened or closed by tension. A stiff floor does not speak in words, yet through these messengers it can prompt a cell to build thicker cables, stronger adhesions, and a more assertive programme of growth and survival, written in the quiet, stubborn syntax of transcription.

## 5.3 Wiring the Circuit

Mechanics can feel like a command, but inside a cell it is handled more like a conversation. A stiff matrix pushes a cell toward tension, yet the cell is not a passive spring. It actively adjusts how hard it pulls, how strongly it attaches, how much internal scaffolding it builds, and how willing it is to change its gene expression. Those adjustments are not isolated responses. They are coupled, wired together into a circuit where cause and effect chase one another in loops.

The temptation is to look for a single "mechanical pathway," a neat line from outside to inside, like a wire carrying current from a switch to a lamp. In reality, the wiring resembles a feedback-controlled system: adhesions alter the cytoskeleton, the cytoskeleton alters the nucleus, and nuclear transcription alters the adhesions again. That looping architecture is why mechanical cues can become decisions rather than momentary reflexes.

The cell's first contact with the physical world is literal contact. Integrins on the membrane bind to proteins in the extracellular matrix, and on the inside they recruit a dense knot of proteins that biologists call a focal adhesion. The name is slightly misleading. These adhesions are not fixed rivets. They are dynamic assemblies whose composition changes under force, and whose lifetime can range from seconds to hours.

Several decades of work, including influential reviews from Benjamin Geiger and colleagues, turned focal adhesions into mechanobiology's central stage. They are crowded with molecules that can sense tension and respond by changing shape, changing binding partners, or recruiting enzymes. Vinculin is a well-known example: it can bind to talin, an integrin-associated protein that itself can unfold under force, exposing additional binding sites. FAK (focal adhesion kinase) is another: it helps transduce adhesion signals into phosphorylation cascades. p130Cas provides yet another angle: it can be stretched, exposing regions that become phosphorylated more readily. The point is not to *memorise* names; it is to recognise a pattern. The adhesion is a load-bearing, force-responsive platform that can translate "pulling harder" into "activating different chemistry."

One can make this idea tactile. Imagine a climbing piton hammered into rock. The harder you load it, the more firmly it wedges, because the geometry shifts under force. Focal adhesions behave like molecular pitons. When the cell pulls against a stiff substrate, the load can stabilise adhesions and promote their growth; the adhesion recruits more structural proteins, thickens the actin cables that attach to it, and becomes better at bearing force. When the substrate is soft, the load is dissipated by deformation of the matrix; adhesions tend not to mature in the same way, and the cell's internal tension stays lower.

This is where the wiring begins: focal adhesions do not merely sense tension, they help *generate* it by anchoring actin filaments so that myosin motors can pull effectively. The dominant engine is actomyosin contractility: actin filaments form networks and bundles, myosin II motors slide filaments past each other, and the cell develops an internal tug-of-war. That tug-of-war is regulated by molecular switches, and one of the most important is the small GTPase RhoA.

RhoA is a protein that toggles between active and inactive states. When active, it promotes actin assembly and stimulates contractility through downstream effectors, notably ROCK (Rho-associated kinase). ROCK increases myosin activity, in part by promoting phosphorylation of myosin light chains and by inhibiting myosin phosphatase. The details can be technical, but the logic is straightforward:

RhoA—ROCK is a dial that turns up cellular pulling. It is also responsive to adhesion state and force, so it participates in feedback. More stable adhesions can promote signalling that activates RhoA; active RhoA increases contractility; stronger contractility further stabilises adhesions on stiff substrates. A cell can therefore "lock in" a high-tension state when the environment can support it.

This loop matters because it helps explain a familiar phenomenon in mechanobiology: mechanical cues can produce hysteresis-like behaviour, where the cell's current state depends on its history. If a cell has already built large adhesions and thick stress fibres, it may maintain tension even if the external stiffness changes slightly. Conversely, a cell that has remained in a low-tension state may resist being pushed into a high-tension programme unless the stiffness cue is strong or sustained. The cell is not simply reading stiffness; it is continuously recalibrating its own mechanical apparatus.

The nuclear messengers discussed earlier, such as YAP/TAZ, fit naturally into this circuit. High tension and spreading favour their nuclear localisation and activity. Once nuclear, YAP/TAZ help activate genes that support the high-tension phenotype: components of the cytoskeleton, adhesion machinery, and sometimes extracellular matrix regulators. In effect, the messenger does not just report tension; it helps build the infrastructure that maintains tension. A stiff matrix can therefore trigger a transcriptional response that makes the cell more capable of generating force on a stiff matrix.

The MRTF-A/SRF axis provides a parallel loop with a different input. MRTF-A's nuclear entry is controlled by actin dynamics, particularly the availability of G-actin monomers. When actin polymerises into filaments, MRTF-A is freed to enter the nucleus and co-activate SRF, driving transcription of cytoskeletal genes. Those genes then increase the cell's capacity to polymerise actin and maintain force-bearing structures. The loop is elegant: the cytoskeleton controls a transcriptional programme that reinforces the cytoskeleton. Under conditions that promote tensional homeostasis—a term used to describe a regulated, relatively stable tension state—the balance of actin polymerisation and retrograde flow, assembly and disassembly, can tune MRTF-A localisation and therefore cytoskeletal gene expression.

At this point, it is worth pausing on a crucial idea: mechanobiological circuits are not solely about force; they are about *control.* A cell wants neither too little tension nor too much. Too little tension and adhesions fail, the cell cannot migrate effectively, and tissues may not be maintained. Too much tension and structures tear, membranes rupture, and signalling can tip into pathological growth or fibrosis-like remodelling. Many cell types appear to regulate themselves into a band of workable tension, adjusting their contractility and adhesions to the mechanical conditions they encounter.

That idea can be made concrete by thinking about how cells behave on different laboratory substrates. On a very stiff dish, a fibroblast typically spreads widely and forms prominent stress fibres. On a soft gel, the same fibroblast spreads less and shows fewer thick bundles. But if one forces strong RhoA signalling on the soft gel, the cell may attempt to generate higher tension than the substrate can support. The result can be unstable adhesions, altered signalling, and sometimes detachment or apoptosis. The environment sets limits, and the circuit adjusts the cell's behaviour within those limits.

The adhesion itself behaves like a local control element. It is not simply "on" or "off." Force can alter the conformation of proteins, the clustering of integrins, and the recruitment of enzymes. Vivien Vogel and Martin Sheetz (2006) gave a memorable framing by distinguishing mechanosensing (detecting force), mechanotransduction (converting it to biochemical signals), and mechanoresponse (changing behaviour). That framing can be helpful as long as it is not taken as strictly sequential. In a living focal adhesion, sensing and response are intertwined. Reinforcement of the adhesion is itself a response that changes the forces experienced by the adhesion, which changes the sensing.

A short historical anecdote captures why the "circuit" view was hard-won. For many years, it was common to treat signalling pathways as if they were separate modules: growth factor signalling here, adhesion signalling there, transcription somewhere else. Part of the power of the early rigidity and durotaxis experiments was that they forced these modules to interact. If a cell can migrate toward stiffness with no chemical gradient, then the adhesion and cytoskeleton must be able to bias protrusion and retraction cycles based on mechanical feedback alone. If a stem cell can drift toward bone-like gene expression on a stiff gel, then transcription must be wired into mechanical state. These were not just new facts; they were violations of tidy compartmentalisation.

The circuit view also clarifies why mechanobiology is deeply relevant to development and disease. Consider tissue fibrosis, where excessive matrix deposition and crosslinking stiffen tissue. Stiffening promotes stronger adhesions and higher RhoA—ROCK activity, which promotes contractility and can activate transcriptional programmes (including YAP/TAZ and MRTF-A/SRF) that support further matrix remodelling. The stiff tissue does not merely reflect damage; it can amplify it by pushing resident cells into a state that produces yet more stiffness. Conversely, in regenerative contexts, a scaffold designed to match the right mechanical properties can bias cells toward constructive behaviours rather than scarring ones.

The same logic can play out in cancer, where many tumours stiffen their microenvironment. A stiffened matrix can support high traction forces and nuclear YAP/TAZ activity, enhancing survival and proliferation programmes. Again, stiffness is not the sole cause of malignancy, but it becomes part of the selective landscape: the circuit makes certain phenotypes easier to maintain on hard ground.

Mechanobiological wiring also depends on geometry. Cells do not merely read stiffness; they read the spatial arrangement of adhesive sites and the shapes they can adopt. Micropatterning experiments, where cells are forced to adhere to islands of defined geometry, show that simply changing cell shape can shift tension distribution and YAP localisation. That means the circuit is sensitive not only to the material properties of the substrate but to how the cell can organise itself on that substrate. In tissues, geometry is rarely uniform: fibres align, pores open and close, neighbouring cells press and pull, and junctions form. Each of these features changes how forces are routed, which changes how the circuit settles.

What, then, is the "signal" in mechanobiology? It is not stiffness alone. It is a composite of traction forces, adhesion dynamics, cytoskeletal architecture, and nuclear mechanotransduction. The cell's response is not simply "more" or "less" of a molecule. It is a reorganisation of a physical system, stabilised by gene expression.

The most useful mental model is to treat the cell like a self-building machine that is constantly running diagnostics on its own structure. It tests the environment by pulling, senses whether adhesions hold, adjusts RhoA—ROCK activity to match resistance, uses YAP/TAZ and MRTF-A/SRF to tune gene expression, and then rebuilds its adhesions and cytoskeleton accordingly. When the environment remains stable, the cell can reach a steady state: a characteristic spread area, characteristic traction forces, characteristic nuclear signalling. When the environment changes, the system is perturbed and must settle again, sometimes into a different basin of behaviour.

This helps explain why mechanical decisions can have a "point of no return" feel. Once transcription has reinforced a particular structural state—thick stress fibres, mature adhesions, a stiffened nucleus—the cell's mechanics are altered, and those altered mechanics affect signalling. Commitment is not only a matter of gene regulatory networks; it is also a matter of physical infrastructure. A cell that has built itself for hard floors will experience the world differently, even if moved to a softer bed, because its own stiffness, adhesions, and nuclear mechanics have changed.

A final, human-scaled example can bring the abstraction back to the body. Consider the heart after injury. The damaged region is often replaced not by functional muscle but by scar tissue. The scar changes the mechanical environment for surviving cells, altering the forces they experience with every beat. Those altered forces feed into the mechanobiological circuit: adhesions, RhoA—ROCK, YAP/TAZ, actin-regulated transcription. The tissue's mechanics and the cells' gene expression become entangled. A scar is not merely a patch; it is a new mechanical regime that can perpetuate the very cellular behaviours that maintain it.

The elegance of this wiring is that it makes physical sense. A cell cannot decide its fate in isolation from the material world because its job is to live in that world. The circuit that links adhesions to cytoskeleton to nucleus is a way of ensuring that what is built inside the cell is appropriate for what is felt outside it. When the world pushes back, the cell learns to push back too, reinforcing itself until the forces at its edges and the genes in its nucleus describe the same reality.

Chapter 6

# Weaving the Matrix: Building Tissues and Organs

*A multicellular organism is far more than a mere aggregation of cells; it is a meticulously constructed edifice held together by self-secreted mortar and molecular bolts. This chapter investigates the physical infrastructure of life, exploring how cells manufacture their own environment and mechanically link to neighbors to form continuous, load-bearing tissues. We discover that the shaping of an embryo-from the folding of the neural tube to the elongation of limbs-is fundamentally a feat of biological engineering driven by coordinated physical forces.*

## 6.1 The Living Scaffold

A body is not only made of cells. If you could remove every cell from a piece of tissue, what would remain would not be empty space. You would be left with a pale, ghostly framework: a mesh of fibers and gels that holds shape, channels forces, stores water, and presents chemical signals in the right places. This is the extracellular matrix, or ECM—"extra-cellular" simply meaning "outside cells." Calling it a scaffold is accurate, but dangerously incomplete, because a scaffold suggests something passive: a rigid stage on which life performs. The ECM behaves more like an adjustable set for an ongoing play, rebuilt and re-tensioned while the actors are still moving.

A useful way to think about the ECM is as a composite material, like reinforced concrete. Composite materials gain their remarkable properties by combining components with different strengths. In concrete, brittle cement is reinforced by steel rods. In tissues, a watery gel is reinforced by protein fibers. The analogy matters because it brings mechanics into the picture from the start: different ingredients are mixed to make the material behave differently under pulling, squeezing, and shearing. But unlike concrete, the ECM is alive in the practical sense: it is continuously made, continuously broken down, and continuously reorganized by cells responding to their local environment.

Even the word "matrix" can mislead. It does not mean a uniform blob. Under the microscope the ECM can look like rope, like foam, like lace, or like a swollen sponge. It can be aligned, as in tendon, where collagen fibers run in parallel bundles built for tension. It can be woven in many directions, as in skin, where the aim is toughness from unpredictable pulls. It can be thin and sheet-like, as in basement membranes that underlie epithelia (the cell layers that line organs). It can be heavily hydrated and slippery, as in cartilage and the vitreous humor of the eye. And in every case, cells are not merely sitting in the ECM; they are mechanically embedded in it, tugging on it, compacting it, sensing how it yields, and changing their own behavior in response.

The major structural workhorse of the ECM is collagen. Collagen molecules are long triple helices that assemble into fibrils and then into thicker fibers. Collagen is strong in tension—it resists being pulled apart—and it is the main reason skin does not tear like wet paper. But collagen is not a single thing; there are many types, and different tissues combine them in different architectures. A thick collagen cable is not just "more" matrix than a thin one; it routes force differently and gives cells a different mechanical world.

If collagen is the rope, elastin is the spring. Elastin is the protein that gives tissues the ability to stretch and recoil. The great elastic arteries that leave the heart expand with every beat and spring back between beats, smoothing blood flow. The lungs inflate and deflate millions of times over a lifetime; elastin helps them behave like a resilient bellows rather than a plastic bag that slowly deforms. Without elastin, tissues might still be strong, but they would lose a crucial element of *reversible* deformation: the ability to store mechanical energy and return to shape.

Then there is the gel-like ground substance, where the story becomes more surprising. Many of the ECM's "soft" components are proteoglycans and glycosaminoglycans—large molecules with long sugar chains that carry negative charges. Those charges attract water and create swelling pressure. The result is not just lubrication but a mechanical strategy: a hydrated gel that resists compression. Cartilage, for example, is poor at resisting tension compared to tendon, but it excels at bearing compressive loads in joints because it is a pressurized, water-rich material reinforced by collagen. In other words, cartilage is a fiber-reinforced hydrogel: squeeze it, and water wants to leave, but the dense network slows that flow, producing a time-dependent resistance. That time dependence is what engineers call *viscoelasticity*—a material that behaves partly like a spring and partly like a thick fluid. Many tissues are viscoelastic, and this matters for biology because cells do not pull once and stop; they pull repeatedly, they crawl, they divide, and they endure rhythms like breathing and heartbeat. A viscoelastic matrix filters these forces, sometimes dissipating them, sometimes storing and returning them.

A concrete example sits at the end of your arm. Skin feels soft, but its mechanical demands are brutal: it must stretch, resist tearing, and self-repair while constantly rubbing against the world. Its ECM is built as a layered composite. In the dermis, collagen provides tensile strength and toughness. Elastin allows recoil. Proteoglycans hold water, giving the tissue fullness and contributing to how forces spread. This composite is arranged in an interlaced pattern so that pulling in one direction recruits fibers oriented in that direction, while pulling in another recruits a different set. That is one reason why surgeons pay attention to "Langer's lines"—directions in the skin associated with collagen alignment. Karl Langer, a nineteenth-century anatomist, studied how skin splits when punctured and mapped these preferred directions. His observations were not merely anatomical trivia; they revealed that the matrix has a grain, like wood, and that its internal architecture shapes how the tissue fails.

The deeper point is that the ECM is not a backdrop. It is a mechanical environment that cells inhabit and actively shape. Cells do not simply *experience* stiffness; they *create* it. When fibroblasts (the archetypal matrix-building cells) are placed in a collagen gel in the laboratory, they pull on the fibers and reorganize them. Over time the gel compacts, stiffens, and becomes anisotropic (stiffer in some directions than others). This is not a curious artifact; it is a miniature version of wound contraction in the body, where fibroblasts and their more contractile relatives, myofibroblasts, tug the matrix inward to close a wound. That tugging does more than move tissue; it changes what kind of cells these fibroblasts become. Under sustained tension and the right biochemical signals, they can adopt a more force-generating phenotype, laying down more collagen and remodeling the matrix further. A feedback loop emerges: force alters cells, altered cells alter matrix, altered matrix changes force.

This kind of feedback is crucial for understanding how a tissue maintains a stable mechanical state over years. Consider a tendon. Each time a muscle pulls, the tendon transmits force to bone. If the tendon were to gradually soften, it would stretch more for the same load, changing the mechanics of movement and increasing injury risk. If it were to gradually stiffen too much, it might become brittle. Tendon cells are not indifferent to load; they live in it. Their matrix-building and matrix-degrading activities respond to mechanical use, helping maintain a workable set point. When the balance fails—after injury, in aging, or in disease—the matrix can drift into a mechanically unhealthy state, such as fibrosis, where excessive collagen and altered architecture make tissue stiff and dysfunctional.

The ECM also acts as a chemical information system, and the mechanical and chemical roles are entangled. Many signaling molecules, including growth factors, do not simply float freely until they bump into a receptor. They are stored, tethered, or presented by matrix components. Some are bound to proteoglycans; others are held in latent complexes that can be activated by mechanical stretching of matrix proteins. This means that remodeling the matrix—cutting fibers, rearranging them, changing crosslinks—can change which signals are available, where, and for how long. In development, this helps explain a puzzle: how the same set of genes can produce exquisitely timed and positioned changes in cell behavior. The matrix is not just a mechanical niche; it is a spatiotemporal organizer for signals.

Developmental biology offers a particularly vivid case of the ECM as a living scaffold. Embryos do not begin as miniature adults with parts waiting to unfold; they are dynamic tissues that change shape by coordinated cell behaviors. During these shape changes, the ECM is produced early and patterned in space. Cells lay it down, attach to it, and pull against it to generate forces that move tissues. A basement membrane, for example, can act as a thin but strong sheet that constrains an epithelial layer, helping it fold in a controlled manner. If you weaken or disrupt that sheet, the same cellular forces can yield a different morphology, like bending a thin metal plate versus a piece of cloth.

The ECM's tunability comes not only from how much of each ingredient is present, but from how they are *organized* and how they are *crosslinked*. Crosslinks are chemical connections between fibers that can make a network stiffer and more resistant to deformation. They are not mere glue; they change how forces percolate through the network. A loose collagen mesh can deform locally around a cell, allowing the cell to pull in and realign nearby fibers. A heavily crosslinked mesh distributes force more widely, making local deformation harder and turning the cell's tug into a more global strain. This alters what a cell "feels" when it pulls, and therefore how it responds.

One of the most arresting demonstrations that cells are sensitive to the mechanical character of their environment came from work in the early 2000s using synthetic gels whose stiffness could be tuned while keeping the chemical composition largely constant. Cells cultured on softer gels tended to spread less and generate less contractile tension, while on stiffer gels they spread more, formed stronger adhesions, and often adopted different fates. The implication was unsettling if you had been trained to think primarily in terms of genes and soluble factors: the *physical* properties of the ECM could steer cell behavior. In a developing organism, where tissues range from very soft (like brain) to much stiffer (like pre-bone), this mechanical range is not a side-effect; it is part of the information landscape.

Yet stiffness is not a single number. Two matrices can have the same measured stiffness and still behave differently to cells, because cells probe their environment locally and dynamically. A fibrous matrix, like collagen, can stiffen as it is strained: pull it harder and more fibers align with the pull, increasing resistance. This is called *strain stiffening*. Many biopolymer networks do this, and it has an important biological consequence: as a cell increases its contractile force, the matrix can respond by becoming effectively stiffer, feeding back onto the cell. In a gel that does not strain-stiffen, the same increase in pulling might not change the resistance in the same way. The matrix is thus not only a material but a mechanical partner in a dance, responding to the forces cells apply.

The ECM's dynamism is perhaps most evident in how it is remodeled. Remodeling is a polite word for controlled destruction. Cells secrete enzymes, such as matrix metalloproteinases (MMPs), that cleave matrix components. This allows cells to migrate through tissues, carve paths, and reshape their surroundings. But degradation is not simply removal; it can generate new fragments with signaling activity, expose hidden binding sites, or release stored growth factors. At the same time, cells secrete new matrix proteins and organize them. In many tissues, this happens continuously at a low level, like maintenance work on a bridge. In growth, repair, and disease, the pace and pattern of remodeling change dramatically.

Wound healing shows how the ECM can switch roles within days. Immediately after injury, the body must stop bleeding, producing a clot—itself a temporary matrix rich in fibrin. Cells migrate into this provisional scaffold. Fibroblasts begin to lay down collagen, initially in a disorganized manner. Over time, the collagen is remodeled: fibers become more aligned, crosslinking changes, and the tissue strengthens. In the best outcome, the matrix gradually approaches the mechanical performance needed for the tissue's function. In scarring, the matrix is strong but often overly stiff and poorly organized for the original task. In fibrosis, this stiffening becomes pathological, compromising organ function. The living scaffold can therefore be repaired, but it can also be rebuilt into something mechanically wrong, with consequences as serious as heart failure or liver dysfunction.

The intimate relationship between cells and matrix also creates a kind of long-range communication system. A cell pulling on the ECM does not only change its own immediate neighborhood. In a fibrous matrix, forces can travel along aligned fibers over surprisingly long distances. This means that one cell's contractility can influence another cell's behavior even without direct contact, simply by mechanically pre-stressing the matrix or reorienting fibers. In a developing tissue, this can help coordinate collective behaviors; in a tumor, it can help malignant cells remodel their surroundings and create tracks that guide invasion. A tumor is not just a lump of abnormal cells; it is an altered tissue whose ECM can become denser, more aligned, and more crosslinked, forming a mechanically distinct microenvironment that encourages further abnormal behavior.

At this point, it becomes natural to ask an apparently naive question: if cells are constantly making and breaking matrix, how does any tissue keep its shape at all? The answer is that stability is actively maintained. A stable tissue is not a static object but a dynamic steady state. Cells continuously sample mechanical stress, respond by adjusting their cytoskeleton and gene expression, and alter matrix turnover accordingly. The point is not to freeze the structure but to keep it within an acceptable mechanical range while allowing renewal and adaptation. This is one reason tissues can be simultaneously robust and changeable: robust because their architecture is reinforced and repaired; changeable because the very processes that maintain them can be redirected by new demands.

The most compelling argument that the ECM is a living scaffold is not philosophical; it is practical. Many modern medical challenges hinge on it. Tissue engineering, for example, initially imagined that if you could build a suitable biodegradable scaffold and seed it with cells, the cells would fill in the rest. This sometimes works, but often the hard part is guiding how cells remodel the scaffold into a functional matrix with the right organization and mechanics. A lump of collagen is not a tendon; a soft gel is not a brain. The difference is in architecture, crosslinking, composition, and the mechanical conversation between cells and their surroundings. Even in the clinic, the success of an implant can depend on how the surrounding tissue responds: whether it integrates, vascularizes, and remodels in a controlled way, or walls it off with a fibrotic capsule.

Perhaps the most human-scale consequence of the living scaffold is aging. Over decades, many matrices become more crosslinked and less easily remodeled. Some crosslinks arise enzymatically; others form through non-enzymatic chemical reactions associated with long-term exposure to sugars. The tissue may become stiffer, but not necessarily stronger in a useful way, because stiffness can come with brittleness and reduced capacity for repair. Skin loses elasticity; arteries stiffen; tendons become more injury-prone. These changes are often experienced as "wear and tear," but they are also shifts in the material logic of the ECM: the composite has been chemically and architecturally altered.

There is a quiet elegance to the idea that bodies are built not only by assembling cells, but by weaving and tuning the material between them. The ECM is not merely what cells secrete; it is what cells *live in*, and what they *talk through.* It is strong and soft at once, solid and fluid over time, stubborn under sudden pull and yielding under slow pressure. It can be reinforced like a ship's rigging, pressurized like a waterbed, laminated like plywood, or stretched like an elastic band, depending on what a tissue must endure. And because cells can rebuild it as they go, the scaffold is never finished—it is continually being edited, tensioned, and renewed, so that living matter can hold its shape while remaining free to change it.

## 6.2 The Molecular Handshake

A sheet of cells looks calm in a textbook diagram: neat polygons tiled like bathroom grout. In real life it is more like a crowd linked arm-in-arm. People shift their weight, someone stumbles, the whole line tightens, loosens, re-grips. A tissue has to do something similar. Cells proliferate, change shape, crawl past one another, and endure constant external stresses—blood pressure, breathing, peristalsis, gravity, the abrasion of skin against clothing. If cells were merely adjacent, a tissue would behave like a pile of coins: easy to scatter, unable to carry tension. Coherence requires mechanical coupling, a way for one cell to *share* its forces with its neighbors.

That sharing happens at cell—cell junctions, specialized molecular machines that link one cell's internal skeleton to another's. The workhorse of force-bearing junctions in many tissues is the adherens junction. It is built from cadherins, a family of adhesion proteins whose name comes from calcium-dependent adherence. Cadherins are transmembrane proteins: part of the molecule reaches outside the cell, part spans the membrane, and a tail sits inside. When two cells meet, cadherins on one cell bind cadherins on the other in the narrow gap between membranes. Many cadherins do this in parallel, forming something like a zipper. Inside the cell, cadherin tails recruit a set of adaptor proteins called catenins, which connect to actin filaments—the dynamic protein cables that, together with myosin motors, generate contractile tension. An adherens junction is therefore not just glue. It is a mechanical bridge: actin in cell A $\rightarrow$ cadherin complex $\rightarrow$ actin in cell B.

The word *handshake* captures an essential feature: the connection is cooperative and informative. In a handshake you do not only make contact; you test firmness, you adjust grip, you sense resistance. A cell does something analogous. It pulls on its neighbors through the junction, and the junction responds by strengthening or weakening. This responsiveness is what makes an epithelium tough without being brittle, able to maintain integrity while still reshaping during growth and repair.

A historical detour helps, because adherens junctions became important long before anyone could name a cadherin. In the late nineteenth and early twentieth centuries, embryologists argued about whether cells were largely independent units whose behavior was dictated internally, or whether their interactions created higher-order organization. The microscope showed boundaries between cells, but it could not reveal how forces were transmitted across those boundaries. As electron microscopy developed in the mid-twentieth century, researchers could finally see specialized junctional structures: dense plaques where membranes approached and intracellular filaments seemed to converge. The tissue did not look like a heap of individual cells; it looked like a mechanically integrated fabric.

Cadherins entered the story later, with biochemical purification and molecular cloning. The key conceptual shift was to treat adhesion as a regulated, mechanosensitive process rather than a static property. A cell can increase or decrease the amount of cadherin at its surface, cluster cadherins into stronger assemblies, and tune how strongly cadherins couple to the actin cytoskeleton. That tunability is crucial: a developing tissue needs junctions that can be remodeled, but an adult tissue needs junctions that can withstand chronic stress.

At the scale of molecules, cadherins behave like a self-assembling zipper whose teeth lock only in the presence of calcium ions. Calcium stabilizes the extracellular cadherin domains into a relatively rigid shape, which allows them to bind their partners across the intercellular gap. Remove calcium, and many cadherin-mediated adhesions fail quickly; cells that were in a cohesive sheet can peel apart. This simple observation, used in countless laboratories, is a reminder that mechanical integrity can depend on chemistry as small as an ion.

The inside of the junction is where mechanics becomes especially vivid. Cadherin tails bind $\beta$-catenin and p120-catenin, and through them $\alpha$-catenin, which is the key adaptor to actin. Actin itself is not a static scaffold. It is constantly polymerizing and depolymerizing, and it is actively pulled by myosin motors. When a cell contracts its cortex—the thin actin-rich layer under the membrane—it generates tension along the cell perimeter. In a tissue, that tension does not stop at the edge of the cell. It can be transmitted through adherens junctions into neighboring cells, which pass it onward, distributing stress across many cell diameters. This is how an epithelial sheet behaves more like a continuous material than a mosaic of individual parts.

A concrete example appears every time you cut yourself. The edges of a shallow wound in skin often pull apart, not because cells have suddenly lost their desire to stick together, but because the surrounding tissue is under tension. Skin is mechanically pre-stressed; collagen fibers and the contractile behavior of cells keep it taut. When the surface is breached, that tension redistributes, and the wound gapes. Healing then demands not only new matrix and new cells, but coordinated force transmission across the epithelial layer as it crawls and closes the gap. Adherens junctions allow the advancing front to move as a collective rather than as a set of isolated, competing cells. If junctions are too weak, cells can scatter; if too strong, the sheet can become rigid and tear instead of flowing smoothly.

This balance between solidity and fluidity is one of the most interesting physical properties of living tissues. A tissue can behave like a solid when it must resist deformation, and like a fluid when it must remodel. Adherens junctions sit at the heart of that duality, because they determine whether forces are shared across the sheet or dissipated in local rearrangements. When junctions are reinforced and well-coupled to actin, tension spreads. When coupling is loosened, cells can more easily slide past one another. Development and repair repeatedly switch between these modes.

How does a junction *know* when to reinforce? One answer is that the junction is built to change its own molecular architecture under load. Around 2010, work on $\alpha$-catenin highlighted a particularly elegant mechanism: tension can change protein conformation in a way that recruits additional components. Under pulling force, $\alpha$-catenin can expose binding sites for vinculin, another adaptor protein that strengthens the connection to actin. Vinculin recruitment is not simply decorative; it can increase the mechanical linkage, allowing the junction to bear higher loads. The junction, in other words, contains a force-activated switch. Pulling makes it *more* able to resist pulling.

This is an instance of mechanotransduction: the conversion of mechanical input into biochemical change. The conversion does not require a dedicated "force receptor" in the way the eye requires photoreceptors. It can arise from the basic physics of macromolecules. Proteins are not rigid; they breathe, fluctuate, and occasionally unfold. A force of only a few piconewtons (a trillionth of a newton) can bias those fluctuations, making one conformation more likely than another. If a binding site is buried in one conformation and exposed in another, force can regulate binding. The cell then reads force as a change in which partners assemble at a junction.

It is tempting to imagine that cadherins themselves are the main load-bearing elements, like rivets. They certainly matter. But the junction's strength is not only the sum of cadherin bonds; it depends on how those bonds are organized and how well they are coupled to the actomyosin network. A cadherin bond that is not linked to actin may hold two membranes together briefly, but it will not transmit the kinds of forces needed to coordinate tissue shape. Conversely, a strong actin network without stable cadherin clustering cannot share tension across cells. The adherens junction is therefore best thought of as an integrated *system*: extracellular binding, intracellular coupling, and a force-generating cytoskeleton that both loads and reinforces the connection.

The system's behavior becomes clearer if you picture a line of people holding hands while each person also holds a stretchy belt around their own waist. The belt stands in for the actomyosin cortex, which can tighten or relax. If everyone tightens their belt while keeping hands linked, tension spreads along the line; the group stiffens. If handholds loosen, tightening the belt simply makes each person tense in isolation; the line no longer functions as a unit. In epithelia, actomyosin contractility can be patterned so that some junctions become more contractile than others, generating directional forces that reshape the sheet. Cadherins and their associated proteins are the couplers that decide whether those forces remain local or become tissue-wide.

There is also a subtlety that makes living junctions different from ordinary fasteners: they are renewable. Individual cadherin bonds form and break, and cadherin clusters are continuously turned over by endocytosis and recycling. This turnover is not necessarily a weakness. It can allow a junction to dissipate damaging stress, like a climbing rope that slides slightly through a belay device instead of snapping. At the same time, turnover creates opportunities for regulation: signaling pathways can increase cadherin delivery to the membrane, stabilize clusters, or alter the actin linkage. The tissue can therefore tune its cohesiveness in response to context—tightening to resist injury, loosening to allow remodeling.

A dramatic example of what happens when this system fails comes from blistering diseases. In some conditions, loss of proper adhesion in the skin causes layers to separate under ordinary mechanical stress, producing fluid-filled blisters. Many such disorders involve desmosomes (another type of junction specialized for intermediate filaments), but the broader lesson applies: everyday life is mechanically demanding, and the integrity of tissues depends on molecular linkages that are usually taken for granted. When adhesion fails, biology does not merely look messy; it becomes physically unlivable.

The same logic applies in cancer, where altered cell—cell adhesion can change how cells move. A carcinoma often begins in an epithelium, where cells are normally tightly coupled. Shifts in cadherin expression and junctional regulation can accompany the transition from a cohesive tumor mass to invasive behavior. It is not that "adhesion molecules cause cancer" in a simple sense, but that mechanical coupling constrains how cells can reorganize. Reduce coupling, and cells can more readily detach and migrate; increase coupling inappropriately, and a tissue may become abnormally stiff or fail to remodel in healthy ways. The molecular handshake influences the social behavior of cells.

Adherens junctions also coordinate with the cell's relationship to the extracellular matrix. A tissue is not built from cell—cell connections alone; it is anchored to and shaped by the matrix outside. When an epithelial sheet adheres to a basement membrane, forces can be transmitted both laterally (through adherens junctions) and vertically (through cell—matrix adhesions). These force pathways interact. Tension at cell—cell junctions can influence how strongly cells adhere to the matrix, and vice versa. A tissue that must resist stretching, like the lining of the gut, relies on this multi-axial force sharing. If one pathway is compromised, the other may be overloaded.

The beauty of the adherens junction is that it makes mechanical continuity out of individual living units without erasing their individuality. Each cell can generate force, sense force, and respond. Yet because the forces are coupled, a local change can propagate. A few cells tightening their cortex can increase tension across many neighbors. A few cells losing adhesion can create a weak spot that concentrates stress. Tissues therefore have a kind of emergent mechanics: properties that belong to the collective, not to any single cell.

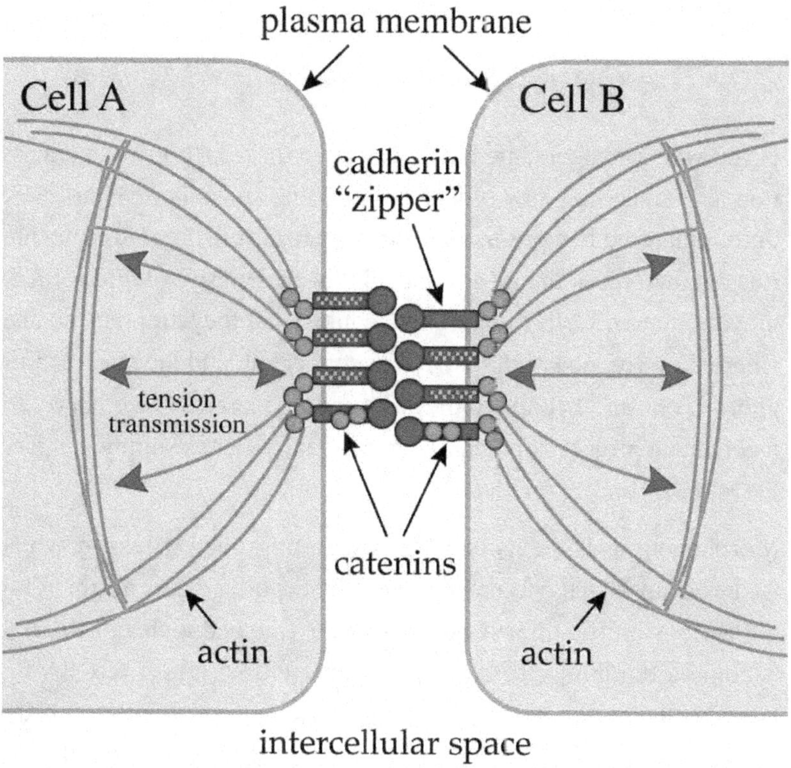


There is a final, almost philosophical payoff to thinking about adherens junctions as a molecular handshake. A handshake is not only a bond; it is an agreement about how forces will be handled. In a tissue, that agreement makes it possible for cells to act collectively without surrendering their ability to respond locally. Cadherins bind across a tiny gap, catenins translate that binding into an internal linkage, actin and myosin supply adjustable tension, and force itself helps decide how strong the connection should be. From the outside we see a smooth surface—skin, gut lining, the delicate wall of a developing organ. Underneath, millions of microscopic handshakes continuously grip, test, reinforce, and release, so that living sheets can be both steadfast and alive to change.

## 6.3 The Origami of Development

If you want to understand how an embryo makes a body, the temptation is to imagine a tiny blueprint unfolding: head here, heart there, limbs emerging like pre-marked tabs. That picture is comforting, and it is also wrong in the most interesting way. Embryos build form by moving matter. Cells do not only change what they are; they change where they are, how tightly they are connected, and the stresses they apply to one another and to the surrounding matrix. The embryo is a mechanical project as much as a genetic one. Genes supply the parts list and the timing; forces supply the sculpting.

Morphogenesis is the name for this sculpting: the process by which tissues change shape to make organs. The word sounds abstract, but the actions are familiar if you have ever watched a sheet of dough become a dumpling, or a piece of paper crease into a box. A flat layer bends into a tube; a long tissue narrows and extends; a cluster hollows out into a cavity. The embryo does not have hands, but it has something just as effective: countless cells that can pull on their own internal scaffolds, grip their neighbors, and attach to extracellular material. Those microscopic pulls can add up to macroscopic folds.

A helpful starting point is to think about where the forces come from. Many of the forces that reshape embryos are generated by the actomyosin cortex: a thin layer under the cell membrane made of actin filaments and myosin motors. Myosin consumes chemical energy (ultimately from food) and converts it into mechanical work, sliding along actin and tightening the network. When this tightening happens uniformly, a cell can round up, like a drawstring bag being pulled closed. When tightening is localized to one side of a cell, it can change shape in a directional way. That directionality is the key to making folds rather than mere shrinkage.

One of the most iconic examples is apical constriction. Many epithelial cells are polarized, meaning they have an apical side facing a lumen or the outside world and a basal side facing underlying tissue. If a group of epithelial cells contracts its apical surface, each cell becomes wedge-shaped: narrow at the top and broader at the bottom. A patch of wedges in an otherwise flat sheet creates curvature, because the sheet has effectively become shorter on one side than the other. Multiply this across thousands of cells, and you can bend a plane into a groove, then into a tube. The process is conceptually simple; the execution in an embryo is exquisite.

A historical anecdote captures how much of this was understood before molecular biology could name the components. In the early twentieth century, embryologists watching sea urchins, frogs, and chickens described invaginations and folds with almost architectural language. They could see tissues buckle inward, but they debated what caused it. Were cells being pushed by neighbors? Were they swelling? Were they actively pulling? With better microscopy and the ability to perturb embryos, the field gradually turned toward active contractility. The modern vocabulary of actin, myosin, and junctions came later, but the essential insight was already present: shape change is not only growth; it is force and geometry.

Neural tube closure is a vivid case where geometry, force, and life-altering consequences meet. Early in vertebrate development, a flat sheet of cells called the neural plate bends and rolls into a tube that will become the brain and spinal cord. Failure of this closure can result in severe birth defects. The mechanics are not a single trick repeated; they involve coordinated behaviors across the tissue: local bending at hinge points, changes in cell shape, rearrangements among cells, and interactions with surrounding tissues and matrices.

Apical constriction plays a central role at hinge points, where the sheet must bend sharply. A key molecular player is a protein called Shroom. Experiments in the 2000s showed that Shroom can drive apical constriction in polarized epithelia, and that it is required for proper bending at hinge regions in the neural plate. The value of naming a molecule here is not mere detail. It shows how genes can specify a mechanical behavior without directly specifying a shape. A gene product does not have to encode a "neural tube"; it can encode a way of generating tension at the right place and time. The tube then emerges as a consequence of physical rules applied to living material.

The tissue does not fold in a vacuum. Adherens junctions, built on cadherins, provide the couplings that let apical constriction become a collective act rather than a set of isolated cell deformations. If one cell constricted apically but did not transmit that tension to its neighbors, it would simply pucker and perhaps detach. When many cells constrict while remaining mechanically linked, the sheet bends coherently. This is where the "molecular handshake" becomes a morphogenetic machine: force generated by actomyosin is routed through junctions into tissue-scale curvature.

There is a second major class of morphogenetic movement that looks less like folding and more like flow: convergent extension. Here a tissue narrows in one direction (converges) and lengthens in the perpendicular direction (extends). A classic example occurs during gastrulation, the dramatic rearrangement early in development when germ layers are formed. Cells intercalate—they slide between one another—so that a wide field becomes a long, narrow strip. The visual analogy is lanes of traffic merging: cars do not vanish, but the pattern changes, and the stream extends.

Intercalation requires cells to make and break junctions, and to do so with directionality. The cell cannot simply loosen all attachments, or it would detach and drift. Instead, it rearranges contacts in a biased way: reducing junctions along one axis while strengthening along another. Planar cell polarity (PCP) signaling provides part of the directional cue. PCP is a set of molecular pathways that give cells a sense of orientation within the plane of a tissue, like a compass embedded in each cell. PCP can polarize actomyosin contractility to specific junctions, often through regulators such as Rho kinase. The mechanical consequence is that certain junctions experience higher tension and are more likely to shrink, pulling cells into new neighbors. Over time, these biased junctional changes can elongate the tissue.

A useful way to keep these processes grounded is to imagine the embryo as a material with competing demands. It must be cohesive enough to transmit forces, but able to remodel without tearing. It must be stiff enough to hold emerging shapes, but soft enough to be reshaped again. Those demands echo the properties of the extracellular matrix described earlier: the matrix can be stiff or compliant, aligned or isotropic, and it can change with remodeling. In development, the ECM is not only a substrate; it is a constraint and a guide. A basement membrane can prevent cells from bulging outward, forcing bending to occur inward instead. A fibrous matrix can provide tracks that orient migration. A hydrated gel can resist compression, altering how an invagination proceeds. When tissues fold, they often do so against a mechanical boundary condition set by matrix and neighboring tissues.

An embryo therefore behaves less like an isolated sheet and more like a connected system of layers. When one layer changes shape, it can drag or compress another. The neural plate, for example, interacts mechanically with adjacent tissues; closure involves not only the bending of the neural epithelium but also forces from surrounding ectoderm and mesoderm. This can be frustrating if one wants a single elegant mechanism. It is also a mark of how biology achieves reliability: multiple partially redundant forces can steer the system toward the same outcome, like several hands guiding a flexible strip into a groove.

The reliability is all the more remarkable because the material is noisy. Cells do not all constrict at exactly the same moment or with the same strength. Junctions fluctuate. Adhesion molecules turn over. The ECM is remodeled. Yet the tissue-level shape change is often smooth and reproducible. Part of the explanation lies in feedback. Tension can stabilize junctions; stretched proteins can recruit more adaptors; actomyosin networks can align with principal stress directions. In physical terms, the tissue can self-organize around force patterns.

One of the most illuminating experimental approaches has been to measure forces directly or infer them from tissue geometry. Laser ablation experiments, for instance, cut a small junction or a patch of cortex in a living embryo and observe how the tissue snaps back. The recoil reveals that the tissue was under tension, and the direction of recoil reveals the direction of that tension. These experiments turn an embryo into a self-reporting mechanical system: cut, and it tells you where it was stressed. They also make a conceptual point that is easy to miss when staring at gene expression maps. Many developmental processes occur in a pre-stressed tissue, like a tent under tension. Changing one cable can shift the whole structure.

Another concrete example is the formation of tubes beyond the nervous system. Many organs use tubular architecture because tubes transport fluids and gases efficiently: blood vessels, gut, airways, kidney nephrons. There are multiple ways to make a tube. Some arise by wrapping and sealing an epithelial sheet; others arise by hollowing out a solid rod of cells (a process called cavitation); others by budding and branching. In each case, mechanics matters. To wrap and seal, a sheet must bend and bring edges into contact without ripping. To hollow out, cells must coordinate adhesion and polarity so that a lumen opens and expands. To branch, localized forces must reshape the epithelium while the surrounding matrix yields or is remodeled to permit outgrowth.

Branching morphogenesis in the lung and kidney provides an especially tangible intersection of genetics, mechanics, and matrix. Signaling molecules pattern where branches should form, but the physical act of branching requires that cells proliferate, migrate, and reorganize while pushing into or pulling against the surrounding ECM. Too stiff a matrix can physically impede branching; too soft a matrix can fail to provide the resistance needed for a stable branch. This is not a metaphorical resistance. Cells generate traction forces through their adhesions; without an opposing load, the same contractile machinery can fail to organize properly, much like trying to climb a rope that is not anchored.

At the level of individual cells, morphogenetic behaviors often reduce to changes in where actomyosin is concentrated and how it is coupled to neighbors. Apical constriction requires a contractile network at the apical surface linked to adherens junctions. Convergent extension requires planar-polarized tension at specific junctions and controlled remodeling of adhesion. Epithelial bending can be driven by differences in tension between apical and basal surfaces, or by differential growth that creates buckling. Even a seemingly simple event like a cell dividing can be mechanically consequential, because mitosis changes cell shape and tension in a neighborhood. Division can locally relax tension or increase it, depending on context, and the orientation of the division plane can contribute to tissue elongation.

The embryo's "origami" is therefore not a single folding pattern but a repertoire of mechanical motifs. What makes them biological is that they are deployed with exquisite spatial and temporal control. A patch of cells constricts apically here but not there; junctions shrink along one axis but not the perpendicular one; matrix is locally softened by enzymes to allow movement; elsewhere it is reinforced to hold shape. The control comes from gene-regulated signaling pathways, but the execution is physical. One can often predict the direction of bending or extension by asking a mechanical question: where is tension highest, where is resistance lowest, and how are forces coupled?

This perspective changes how developmental defects are understood. A genetic mutation may cause a defect not because a particular "structure gene" is missing, but because a mechanical process fails. If apical constriction is reduced, a fold may not initiate. If junctions cannot bear tension, a sheet may tear or fail to transmit force. If the matrix is not remodeled appropriately, a budding organ may be constrained. These failures can cascade, because morphogenesis is sequential in time: one shape change creates the boundary conditions for the next. Even subtle mechanical alterations can therefore lead to major anatomical outcomes.

There is a particular poignancy to this, because it emphasizes that life is built from matter obeying the same laws as any other material, yet achieving forms that no engineer could easily assemble at that scale. The embryo does not have bolts or welds. It has reversible molecular bonds that can be strengthened by force, fibers that align under stress, gels that swell with water, and cells that act as both sensors and motors. Development is not only a story of information; it is a story of work, of energy spent to push, pull, and hold.

In the end, the most astonishing thing about morphogenesis may be its tact. A growing embryo must reshape itself while preserving function in the tissues that are already working. Even early on, cells must maintain barriers, manage fluid pressures, and keep compartments separate. Folding a sheet into a tube is risky: it creates seams, introduces curvature stresses, and demands precise coordination so that the right surfaces end up facing the right way. That it happens at all is a triumph of collective mechanics. That it happens reliably, across millions of births, is an even greater one.

A living tissue can fold because its components are not only connected but responsive. Junctions tighten when pulled. Cytoskeletal networks thicken where stress concentrates. Matrices yield when enzymatically softened and stiffen when crosslinked. Local contractions become global shape because forces are transmitted rather than trapped. The embryo's shapes are therefore not imposed from above; they emerge from countless local decisions—where to grip, where to pull, where to let go—made by cells that are, in a very literal sense, reading and writing force as they build a body.

Chapter 7

# Closing the Breach: The Mechanics of Healing

*When the skin's integrity is breached, a sophisticated mechanical symphony begins, transforming soft cells into high-tension biological sutures. This chapter traces the physical journey of repair, from the initial shock of lost tension to the mobilization of cellular contractors that navigate invisible stiffness gradients to stitch the body back together.*

## 7.1 The Mechanical Void

A cut is not only a tear in a surface. It is a sudden rearrangement of forces.

Before anything goes wrong, most living tissues are quietly *prestressed.* Skin, for example, sits under a background tension created by its collagen weave, the pull of underlying muscles, and the internal pressure of fluids. Surgeons have long learned to respect this invisible tug-of-war. In the nineteenth century, Karl Langer mapped what are now called *Langer's lines* by poking round holes in cadaver skin and watching them spring into ovals. The ovals' long axis revealed the direction of greatest tension, and the lesson remains practical: cut along the lines and a wound gapes less; cut across them and it yawns open. Even without the language of mechanobiology, Langer was reading a mechanical map written into the body.

Injury erases that map locally, and the erasure is abrupt. The simplest way to picture it is to imagine a stretched fabric that has been snipped: the threads at the edge recoil and the tension redistributes outward. Something similar happens in epithelia, the sheets of cells that line surfaces such as skin, gut, and cornea. Those cells are not passive tiles; they are contractile, like tiny musclelets, with a thin "corset" of actin and myosin under their membrane. In an intact sheet, each cell pulls on its neighbors, and the neighbors pull back, creating a balanced field of forces.

When a wound opens, that balance is broken. Forces that were once shared across a continuous sheet find empty space. The first mechanical event is not migration or growth; it is *loss*. A gap appears where tension used to be borne, and the tissue around it enters a mechanically unusual state: slack in one sense (because the continuous load path has been interrupted) and yet sharply stressed in another (because forces concentrate at edges and corners). Engineers would call this a problem of boundary conditions. Cells experience it as a shock.

One reason the shock matters is that cells constantly use force as information. They do not wait for a chemical messenger to arrive and politely announce that damage has occurred. They feel the absence of resistance, the changed pull from neighbors, and the altered stiffness of what lies beneath them. A wound is, in this way, a *mechanical void*: not a vacuum of matter, but a void in the normal architecture of tension and support.

The body's first priority is to stop bleeding, yet even that familiar emergency response has a mechanical side that is easy to miss when described only in biochemical terms. Blood coagulation is often taught as a cascade of proteins that ends with fibrin, a polymer that forms a mesh. True, but mesh is the wrong mental image if it suggests something delicate and limp. A forming clot is a *scaffold*, and scaffolds are mechanical objects. They bear load, resist deformation, and provide a surface that cells can grip.

To appreciate how quickly the mechanical environment changes, keep the timeline tight. Within seconds to minutes, the wound space fills with fluid and cells, pressure shifts, and the cut edges recoil. Platelets adhere, become activated, and bind fibrinogen; fibrin strands begin to span the breach. The clot is not merely a plug; it is an improvised bridge across which force can once again be transmitted. Meanwhile, platelets tug on the fibrin network. They do not tug for aesthetic reasons. Their contraction compacts the clot, squeezes out serum, and stiffens the scaffold. The result is an early, temporary material that is often markedly stiffer than the uninjured soft tissue it replaces. In mechanical terms, the wound does not remain a "hole" for long; it becomes a region with new stiffness, new topology, and new patterns of strain.

That stiffness has consequences. Many cells behave differently on a soft mattress than on a hard floor. Fibroblasts, for example, spread more and form stronger adhesions on stiff substrates. Immune cells crawl and squeeze through porous matrices in ways that depend on pore size and compliance. Endothelial cells that later build new blood vessels respond to tension and shear. Even epithelial cells at the wound edge reorganize their internal scaffolding in response to the altered mechanical constraints. The clot therefore acts like an emergency architecture project: fast, messy, and structurally decisive.

It is tempting to imagine that the most important signal is the presence of fibrin itself, as if cells recognize it chemically and then proceed. Yet the physics is at least as informative as the chemistry.

A fibrin network provides *ligands* (binding sites) for integrins, the cell's main "hands" for gripping the extracellular matrix. But it also provides *resistance* when the cell pulls. Integrins are not just hooks; they are part of a sensory system that reads how much the outside world yields. When a cell contracts its actin cytoskeleton against an anchored integrin, the degree of movement and the lifetime of the adhesion both depend on external stiffness. In plain terms, cells test their surroundings by tugging. The clot gives them something to tug *against*.

Around the wound, there is another, less obvious mechanical event: a rapid reshaping of tension not only at the edge but across a wider neighborhood of the tissue. In living epithelia, measurements using delicate micromechanical approaches suggest that cortical tension can drop broadly within tens of seconds after damage, as if the sheet briefly "lets go." That sounds counterintuitive. Shouldn't cells tighten up to protect themselves? But a short-lived release may be a way to prevent tearing from propagating, much like a safety clutch that disengages under overload. After that release, tension is actively restored. What matters for healing is not a single force value but a dynamic pattern: tension falls, then returns, and the return can propagate like a wave. In such a picture, the wound is not merely a static gap; it is a disturbance that travels through the mechanical field of the tissue.

How can a tissue coordinate a tension response so quickly? Part of the answer is calcium. Calcium ions are a universal alarm bell in cells. They surge when membranes are breached or when mechanosensitive pathways open, and they can spread from cell to cell through junctions. Calcium can rapidly alter actomyosin contractility by activating myosin light-chain kinase and other regulators, essentially turning up or down the cell's internal motors. In some systems, the mechanical "quiet" of healthy epithelium can be broken by a flash of calcium, followed by organized flows of actin and myosin that move across multiple cells. The choreography is striking: instead of each cell acting alone, the sheet behaves like a coupled material whose components are alive.

The most immediate mechanical puzzle of a wound is this: if a gap has opened, what exactly tells nearby cells where the gap is? Chemical gradients help, but they take time to form and can be blurred by fluid flow. Mechanics offers a faster cue. A wound edge changes the geometry of force transmission. Cells adjacent to the gap lose a neighbor on one side; the pull on their junctions becomes asymmetric. Their focal adhesions, where they grip the underlying matrix, also experience altered loading. As these cells contract, they feel less resistance toward the open space and more resistance toward intact tissue. That asymmetry is, effectively, a compass.

Even the simplest everyday experience contains a version of this. If you are pulling a heavy tablecloth taut and someone suddenly cuts a slit, your hands immediately feel the slack and the shift in tension long before you see the exact shape of the cut. Cells have no eyes, but they have hands and muscles at the nanoscale. A wound announces itself as a directional change in resistance.

The clot modifies that announcement. Initially, the gap is a free edge; later, it is a span of fibrin. That transition changes what "toward the wound" feels like. Early on, pulling toward the gap is like pulling into air. Minutes later, it is like pulling into a provisional gel that can be deformed, densified, and remodeled. The wound becomes, mechanically, a place where traction produces visible consequences: fibers align, pores enlarge or collapse, stresses concentrate at anchor points. For a cell that reads its environment through force, the clot is a newly legible surface.

A concrete example helps. Consider a superficial cut in skin. If it is small, it bleeds briefly, then a scab forms. We often treat the scab as a cosmetic nuisance, but it is an engineering solution: a dried, stiffened cap that protects the wound and helps maintain a stable environment beneath. Under that cap, the fibrin-rich clot is hydrated and elastic rather than brittle. It anchors to the wound margins and can transmit forces generated by cells as they crawl and contract. If you pick off a scab early, you are not only removing a chemical barrier; you are removing a mechanical brace and disrupting the stresses that cells have begun to organize around. The result is often renewed bleeding and delayed closure, not because the body has forgotten the steps, but because you have changed the boundary conditions.

Historical surgeons, long before microscopes revealed fibroblasts and immune cells, learned the value of mechanical stabilization. Ambroise Paré, the sixteenth-century French barber-surgeon, is remembered for abandoning the practice of pouring boiling oil into gunshot wounds and for improving ligatures. In the background of those advances lay a practical appreciation: a wound that is approximated, supported, and kept from gaping tends to do better. Today we can translate that folk wisdom into mechanobiology. Bringing wound edges together reduces the mechanical void, decreases the length of free boundary, and limits the region in which cells experience extreme gradients of stress and strain.

The mechanical void is not just a feature of surface wounds. In internal tissues, a tear or infarct similarly disrupts the load-bearing network. Heart muscle after a myocardial infarction offers a dramatic example. A portion of tissue dies and loses contractile function; the remaining tissue must bear altered stresses. Over time, the ventricle can dilate and remodel, with scar tissue forming that is mechanically different from muscle. The later story involves chronic changes, but the first story is immediate: a rupture in the mechanical fabric forces the surviving cells to operate under a new pattern of loads. Those loads are not innocent. They can activate mechanotransduction pathways that change gene expression, metabolism, and cell fate.

Back at the scale of a small skin wound, the clot's stiffness is not fixed. Fibrin networks are *viscoelastic*: they behave partly like a solid and partly like a fluid. Under a sudden pull, they resist; over time, they can relax and creep. Platelets contract, compacting the network, but enzymes begin to cut and remodel it. Fluid flows through it. Cells invade it and pull on it. The result is that stiffness is experienced not as a single number but as a time-dependent property. If a cell tugs quickly, it may feel a stiff response; if it maintains tension, the matrix may yield. This matters because cells can adjust their own pulling speed and pattern. Mechanosensing is therefore not only about "how stiff is it?" but also "how does it respond when I pull like *this*?"

Another subtlety is that the clot is heterogeneous. Near the wound edge, fibrin may be denser and more cross-linked; deeper regions may be looser. Red blood cells, platelets, and trapped proteins create a composite material, not a uniform gel. A cell crawling into the clot experiences a patchwork of stiffness and adhesion. That patchwork can guide movement and organization in the absence of a clear chemical gradient. In many contexts, cells align along fibers that have been stretched. A fibrin network under tension can become anisotropic, with fibers oriented in preferred directions, and cells tend to align with those directions because it is mechanically efficient: pulling along a taut fiber transmits force farther than pulling across slack ones.

This is one reason wounds often close with a characteristic geometry. The mechanical landscape around an oval tear is different from that around a jagged laceration. Corners concentrate stress and can become foci for cellular activity. Smooth edges distribute stress more evenly. Clinically, ragged wounds are more difficult to approximate and often heal with more scarring. Part of that difference is contamination and tissue loss, but part is mechanical: irregular boundaries create complex strain fields, and cells responding to those fields may deposit extracellular matrix in disorganized ways.

The immediate mechanical aftermath of injury also involves fluid pressure. When vessels are cut, local pressure drops; fluid leaks into interstitial spaces; swelling can occur. Edema is often described as an inflammatory symptom, but it is also a mechanical change that stretches tissues and raises interstitial pressure. Stretch alters how cells adhere, how they open ion channels, and how they secrete matrix. Pressure and tension are not separate; they trade off. Inflate a balloon and the surface tension rises. Swell tissue and the tensile state of collagen and cells changes. In a wound, pressure changes can therefore help set the mechanical context in which repair begins.

At the microscopic level, the boundary between "inside" and "outside" is also under mechanical threat. Cells at a wound edge face exposed membrane and disrupted junctions. They must rapidly prevent further tearing and restore a seal. One protective response is to assemble contractile structures near the edge, tightening like a drawstring. Even before a stable contractile ring forms, there can be rapid actomyosin remodeling in the surrounding cells, as if the tissue is gathering its strength. This is not the slower work of building new tissue; it is the quick work of restoring mechanical continuity so that slower processes can proceed on stable ground.

The clot's role as a mechanical bridge becomes especially vivid when considering what happens if clot formation is impaired. In bleeding disorders, or in patients on strong anticoagulants, the early scaffold may be weak or delayed. The wound then remains, for longer, a genuine mechanical void: a gap filled with fluid rather than a load-bearing gel. Cells attempting to migrate encounter a substrate that does not provide stable traction. Forces produced by the tissue may simply deform the fluid-filled space without generating useful movement or alignment. The familiar clinical outcome is prolonged bleeding and delayed healing, but behind that lies a mechanistic explanation: without an early scaffold, the tissue cannot re-establish a workable mechanical environment.

Conversely, an overly rigid or persistent early scaffold can also be problematic. If the provisional matrix is too stiff, it can drive cells to generate higher forces and to adopt more contractile phenotypes. In the long run, that tendency can contribute to excessive scarring and fibrosis, where tissue becomes abnormally stiff and function is compromised. The body therefore faces a balancing act: build a scaffold stiff enough to stabilize and guide repair, yet temporary and compliant enough to be remodeled into something that resembles the original tissue rather than a permanent patch.

It is worth pausing on the emotional dimension of this, because it explains why mechanobiology matters beyond the laboratory. A child's scraped knee and an elderly person's chronic ulcer are separated by more than age. In many chronic wounds, the mechanical environment is deranged: swelling, poor perfusion, altered extracellular matrix, and repeated mechanical insult from pressure or shear. The wound is not allowed to exit the mechanical void state. It remains trapped in a loop of breakdown and incomplete stabilization. Treatments that seem mundane—offloading pressure with special footwear, compressing a swollen leg, debriding a stiff necrotic cap—are, in effect, attempts to repair the mechanics so that biology can proceed.

A wound, then, is not simply a site where "healing factors" arrive. It is a region where the rules of force have changed. The sudden loss of pre-existing tension jolts cells out of their steady state, and the rapid construction of a fibrin-and-platelet scaffold replaces emptiness with a new, tactile world. Cells at the margin begin to pull, not only to move but to *measure*. Each tug tests whether the breach is still empty or whether it has become something that can be gripped, reshaped, and, eventually, rebuilt into living tissue.

## 7.2 Living Sutures

A wound is a gap in tissue, but it is also a gap in *load-bearing*. The clot can bridge that gap temporarily, yet it cannot do the longer job of pulling slack edges into contact and stabilising the new tissue as it forms. That work falls to cells that behave, in effect, like contractors: they arrive, attach, pull, and then decide whether to pack up and leave or stay and keep tightening. The most consequential of these contractors is the myofibroblast, a cell that looks ordinary under the microscope until one asks what it can do to a piece of connective tissue.

Fibroblasts are the baseline workforce of connective tissue. In uninjured skin they live among collagen fibres and other extracellular matrix (the meshwork of proteins and sugars outside cells that gives tissue its structure). Fibroblasts make matrix, repair small wear-and-tear, and talk constantly to their surroundings through integrins, the receptor proteins that anchor a cell to the matrix and transmit force in both directions. In that everyday state, a fibroblast is active but not especially muscular.

After injury, many fibroblasts change character. They spread more, develop thicker internal cables of actin (a structural protein) and myosin (a motor protein), and begin to contract the matrix in a sustained way. Once they also express a particular form of actin called $\alpha$-smooth muscle actin ($\alpha$-SMA), they earn a new name: *myofibroblasts*. The prefix *myo-* means muscle, and it is not poetic licence. These cells acquire a contractile apparatus closer to that of smooth muscle than to a typical connective-tissue cell.

This transformation has a chemical side-signals like transforming growth factor beta (TGF-$\beta$) are famous for pushing fibroblasts toward a myofibroblast state-but it also has a mechanical side that is easy to state and hard to ignore: a fibroblast becomes a myofibroblast when it senses that it is needed to bear force. A loose, soft environment encourages a loose, soft cell. A taut, resistant environment invites a taut, force-generating cell.

One reason mechanics matters so much here is that the myofibroblast is not simply a cell that *makes* a scar. It is a cell that *pulls* on the wound and, by pulling, changes the entire landscape of tension for every other cell trying to rebuild tissue. If the clot provided a temporary bridge, myofibroblasts function more like adjustable cables, tightening the structure as new material is laid down.

The first key idea is that myofibroblasts do not pull in the abstract. They pull on collagen and other matrix fibres through a specialised interface called a *focal adhesion*. Despite the name, focal adhesions are not tiny points; they are complex molecular assemblies, packed with integrins and scaffolding proteins, that link extracellular fibres to the cell's internal actin cytoskeleton. They are both rivets and sensors. When a cell contracts its actomyosin network, the force is transmitted through these adhesions into the matrix. If the matrix is stiff enough, the adhesion grows and strengthens, recruiting more proteins, stabilising the connection, and allowing the cell to generate still more tension. If the matrix is soft and yields, the adhesion remains small and transient. This is mechanotransduction in its most tangible form: force changing biochemistry, which changes structure, which changes force.

The second key idea is that myofibroblasts are not merely acting *on* the matrix; they are acting *through* it. Collagen fibres form a connected network. Pulling on one region transmits stress to another, aligning fibres and tightening slack. A myofibroblast embedded in granulation tissue (the pink, fleshy tissue that fills many wounds) can reorganise collagen over distances much larger than its own body. In doing so, it helps convert a disordered, watery provisional matrix into something that can bear load.

A concrete, everyday example makes the mechanics intuitive. Consider a deep scrape on the shin. In the first week it looks open and soft, with a tender bed that bleeds easily. Later, as it begins to close, the surrounding skin can feel tight, and the wound surface often looks puckered. That puckering is not just drying. It reflects a genuine mechanical contraction of the tissue. Myofibroblasts are generating tension, pulling collagen into alignment and drawing the wound edges inward. If the wound is large, the contraction can be strong enough to restrict movement around a joint. In severe burns, where huge areas of skin are lost, contraction can produce debilitating *contractures*—permanent tightening that bends fingers, elbows, or neck. Clinicians treat contractures with splints and physical therapy, which are, in their own way, mechanical negotiations with myofibroblast-generated forces.

The phrase *living sutures* is not metaphorical window-dressing. Sutures work because they carry tension between wound edges and keep them approximated while tissue rebuilds. Myofibroblasts accomplish something similar from the inside. They fasten themselves to the matrix and apply a steady pull, day after day, as collagen is deposited and cross-linked around them. The result is not a neat thread but a dynamic, distributed network of cellular tethers.

A historical clue that wounds could be mechanically *closed* rather than merely covered comes from the evolution of surgical technique. Long before antibiotics and before anyone had seen a fibroblast, surgeons knew that bringing edges together changed outcomes. The development of reliable suturing and the disciplined practice of wound closure were not only about stopping bleeding; they were about controlling tension. When Joseph Lister's antiseptic methods took hold in the late nineteenth century, cleaner surgery allowed more confident closure, which in turn changed the mechanical environment in which wounds healed. The modern concept of "tension-free closure" echoes a deep mechanobiological truth: excessive tension does not simply threaten to pull a wound apart; it can steer the biology of repair toward fibrosis.

This creates an apparent paradox. If myofibroblasts are needed to close a wound, why fear tension? The resolution lies in distinguishing *productive* tension from *pathological* tension. A temporary increase in contractile force can help restore continuity. Prolonged, excessive force can maintain myofibroblasts in their contractile state and encourage the deposition of dense, disorganised collagen. In other words, tension is both tool and trap.

Boris Hinz and colleagues made this relationship vivid in experiments around 2001 by physically manipulating wound tension in animal models. When healing tissue was mechanically loaded (for example by splinting in ways that prevented contraction and increased tension), myofibroblast markers appeared earlier and persisted longer. When tension was released, myofibroblast features waned over days in a stepwise manner. Those results are striking because they make myofibroblast biology feel less like a fixed genetic programme and more like a decision continuously renegotiated with the mechanical environment. Cells were not simply instructed by a one-time chemical message; they were checking whether their work was still required.

How does a cell decide that? Part of the decision is structural. To become a myofibroblast, a fibroblast builds thicker actin stress fibres and equips them with $\alpha$-SMA, which increases the fibres' ability to generate and sustain tension. It also develops more robust focal adhesions, often called *supermature* focal adhesions in this context, which can transmit higher forces. The whole cell becomes a better winch.

But the decision is also feedback-driven. When a myofibroblast contracts, it can activate signals stored in the matrix. TGF-$\beta$, one of the most potent pro-fibrotic cytokines, is often held in a latent (inactive) form bound to extracellular proteins. In 2007, Pierre-Jean Wipff, Daniel Rifkin and colleagues showed that myofibroblast contraction itself can mechanically activate latent TGF-$\beta$, largely without needing enzymes to cut it free. The logic is beautifully mechanical: the cell pulls on the matrix; the matrix resists; the stored signal is strained into an active state; the active signal then reinforces the myofibroblast programme. This is a biochemical loop powered by force. It also explains why stiff environments can be self-amplifying: stiffness promotes contraction; contraction activates TGF-$\beta$; TGF-$\beta$ promotes more matrix deposition and contractility, which increases stiffness further.

That loop helps clarify a familiar clinical observation: scars often feel stiffer than the surrounding skin, and stiff scars can continue to remodel for months. A scar is not only extra collagen; it is collagen placed and tensioned in a way that changes the mechanical baseline. If the mechanical baseline stays high, myofibroblasts may linger. If it relaxes, they are more likely to disappear, often through programmed cell death (apoptosis) or reversion to a less contractile state. A good scar is, in some sense, one that knows when to stop being reinforced.

The tissue that hosts myofibroblasts is not static while this decision unfolds. Granulation tissue is wet and cellular, rich in provisional matrix proteins like fibronectin as well as collagen. It is also mechanically evolving: fibres become more aligned, cross-links increase, water content changes, and the whole construct becomes more load-bearing. That changing material feeds back on the cells. A myofibroblast pulling on a loose network spends force deforming the network. A myofibroblast pulling on a dense network feels more resistance, which stabilises its adhesions and encourages it to maintain its contractile machinery. The wound is therefore a moving target: cells and matrix co-construct each other.

There is another layer of complexity: wound closure is not performed by myofibroblasts alone, and the most rapid early closure in epithelia can involve actomyosin structures in the epithelial sheet itself-contractile cables that tighten around the edge like a drawstring. In some contexts, a "purse-string" mechanism can even operate at the level of a single wounded cell, assembling a contractile ring that cinches the membrane closed. Yet the deeper, longer-range contraction that reduces wound area over days is strongly associated with myofibroblasts and the collagenous matrix they remodel. It is as if one set of living sutures is thrown quickly to stabilise an edge, while another set is installed more slowly to carry load for the long term.

The most revealing evidence that myofibroblasts truly *generate force* comes from direct measurements. In laboratory settings, fibroblasts cultured in collagen gels can be observed to shrink the gel over time, a simplified echo of wound contraction. When myofibroblast differentiation is promoted, the contraction is stronger. When integrin function is blocked, or when the cytoskeleton is disrupted, contraction weakens. Engineers measure traction forces on flexible substrates; cell biologists watch focal adhesions assemble and disassemble under the microscope. Different tools, same conclusion: a myofibroblast is a force-producing machine coupled to an external scaffold.

That coupling is worth seeing, because it is the physical heart of the phenomenon.

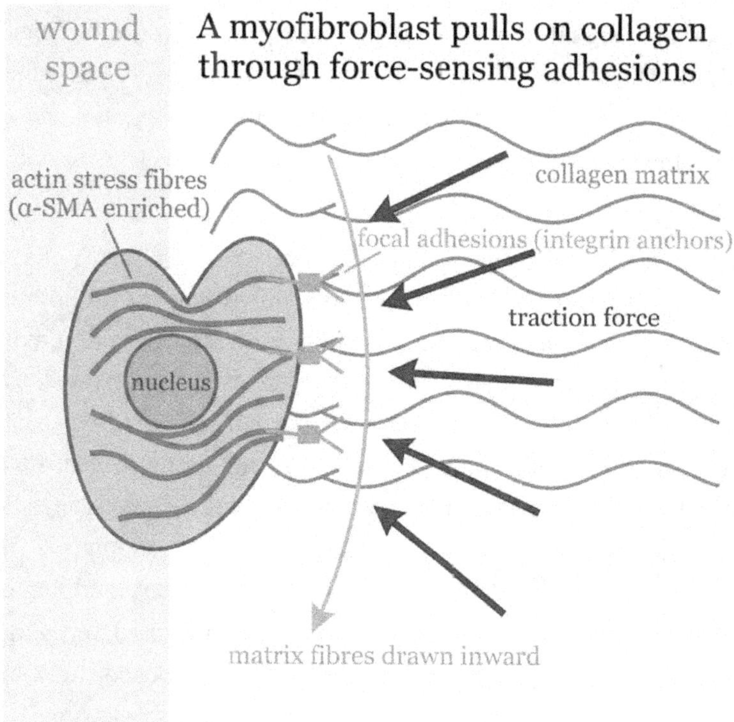


The diagram hides a crucial point in plain sight: the cell's force is not applied to a rigid object. Collagen is deformable, and it is also remodellable. When a myofibroblast pulls repeatedly, it does more than stretch fibres. It can slide fibres relative to one another, align them, and encourage new cross-links to form. Over time, the tissue "learns" the direction of pull, because the matrix becomes organised along the lines of stress. That is one reason scars often have a preferred orientation and why the direction of tension in a wound can shape the final architecture.

Myofibroblasts also have a social life. They communicate with immune cells, which can either resolve inflammation or keep it simmering. They interact with endothelial cells sprouting vessels. They respond to signals from the epithelium closing over the surface. But even these interactions are coloured by mechanics. A stiff, contractile granulation bed sends different cues than a soft one, and cells often interpret stiffness as a proxy for "how mature is this tissue?" or "how much force must I contribute?"

The most clinically charged implication is that treatments can steer healing by steering mechanics. A splint placed to immobilise a wound does not only prevent the patient from reopening it; it reduces cyclic strain that would otherwise keep fibroblasts activated. Compression dressings for venous ulcers are not merely about pushing fluid out; they change interstitial pressure and tissue stretch, altering the mechanical signals that fibroblasts experience. Silicone sheets and massage for hypertrophic scars are, in part, attempts to modulate tension and hydration in the scar environment, nudging cells toward a less contractile, less fibrotic equilibrium. Even surgical choices-where an incision is placed, how it is closed, how much tension remains-can influence how many myofibroblasts appear and how long they persist.

There is a quiet moral in the biology. Myofibroblasts are often cast as villains because they are associated with scarring and fibrosis. Yet they are indispensable in many wounds. Without their contractile labour, large defects would remain open longer, inviting infection and loss of function. The danger arises when a normal emergency workforce becomes a permanent occupation. In that case, the body pays for stability with stiffness.

At its best, wound contraction is a temporary act of architecture: cells assemble a force-transmitting network, tighten it until continuity is restored, and then dismantle the machinery. The body is left with a repaired surface and a connective tissue strong enough to bear everyday loads. If you press a fingertip into a mature scar and feel it resist, you are feeling the outcome of many microscopic tugs, each transmitted through focal adhesions, each resisted by collagen, each shaping the material that now shapes you back.

## 7.3 Navigating by Touch

A healing wound is crowded: immune cells arrive, epithelial sheets creep forward, fibroblasts thread into the provisional matrix, and new vessels begin to sprout. In the middle of this bustle sits a deceptively simple navigation problem. How does a cell, a few tens of micrometres wide and blind to anything beyond its immediate contacts, find the injury and move in the right direction?

Chemical signals do a great deal of the guiding. Damaged cells release molecules that attract immune cells; gradients of growth factors and lipids can point migrating cells toward the defect. Yet wounds are messy chemical landscapes. Fluid flow can wash molecules away. Proteases can chop them up. Diffusion blurs sharp gradients. Meanwhile, the physical consequences of injury are immediate and hard to conceal: a tear changes tension, a clot changes stiffness, and a contracting wound changes strain. Cells are exquisitely sensitive to these physical cues, and they use them as a kind of tactile geography.

The key term is *durotaxis*, from the Latin *durus* for hard and the Greek *taxis* for arrangement or directed movement. Durotaxis means migration up a stiffness gradient: cells tend to crawl from softer regions toward stiffer ones. That preference is not a philosophical stance; it falls out of how cells grip their world.

A migrating cell moves by repeating a cycle. It extends a protrusion at its front edge (often a thin sheet called a lamellipodium), forms new adhesions to the substrate, pulls its cell body forward using actomyosin contraction, and releases adhesions at the rear. Each step depends on force. Adhesions are not merely glue; they are mechanosensitive structures that strengthen when pulled. A cell that "leans" into a stiff region can build robust adhesions and gain traction. A cell that leans into a soft region pulls, but the substrate yields, adhesions stay small and fleeting, and the forward progress is less stable. Over time, tiny differences in adhesion stability and contractile efficiency bias the whole cell's movement toward stiffness.

The remarkable thing is that a cell does not need to measure stiffness with a ruler. It probes stiffness by tugging. Clare Waterman's lab and others popularised a vivid phrasing for this behaviour in the early 2010s: cells *tug* on their environment through fluctuating traction forces. Like a person testing a tent peg by giving it a quick pull, a cell applies local, pulsatile forces through integrins and reads the mechanical response. If the peg holds (a stiff matrix), the cell commits; if it wobbles (a compliant matrix), the cell is more likely to try elsewhere. This is mechanosensing rendered into an everyday action.

In the context of a wound, stiffness gradients are almost unavoidable. The clot is a distinct material inserted into soft tissue. As fibroblasts deposit collagen and myofibroblasts tighten it, the centre of the wound often becomes stiffer than the surrounding uninjured matrix, at least temporarily. Even at the earliest stages, the edge of a wound is mechanically special: it is a boundary where cells lose neighbour contacts on one side and experience altered tension. Add swelling, changes in interstitial pressure, and the evolving architecture of fibrin fibres, and a wound becomes a complex landscape of soft and hard, slack and taut.

A useful image is a mattress with a stiff board slipped under one corner. Without seeing it, you can find the board by pressing with your hand. Cells do something similar at microscopic scale. Their "hand" is a focal adhesion; their "press" is actomyosin contraction;

their "sense" is the molecular response of the adhesion and the cytoskeleton to load.

Wounds, however, are not navigated by solitary cells performing private experiments. Much of repair depends on *collective migration*, in which cells move as groups while maintaining contacts with neighbours. Epithelia provide the clearest example. When a sheet of epithelial cells migrates to cover a gap, it behaves less like a collection of independent hikers and more like a coordinated crowd. The sheet maintains junctions, shares forces, and transmits mechanical information over many cell diameters.

Collective migration introduces a division of labour. Cells at the front edge often act as *leader* cells, extending prominent protrusions, generating strong traction, and exploring the substrate. Cells behind them become *followers*, maintaining cohesion and transmitting force through cell-cell junctions. Leaders pull on the matrix; followers pull on leaders and on each other. If leaders turn, the sheet turns. This arrangement is mechanically efficient: instead of every cell needing a clear external cue, the group can integrate information from the leading edge and propagate a coordinated response.

Mechanical cues are well suited to this collective mode because they are naturally shared. If a leader cell gains traction on a stiff patch and pulls, tension propagates through adherens junctions (the cadherin-based molecular rivets between cells). Neighbouring cells feel the change and polarise accordingly. A chemical gradient, by contrast, must be sensed individually, and noise in sensing can cause cells to disagree. Mechanics offers a kind of built-in consensus mechanism: connected cells are forced to negotiate a common deformation.

It helps to picture what such a tactile gradient might look like in a wound bed.

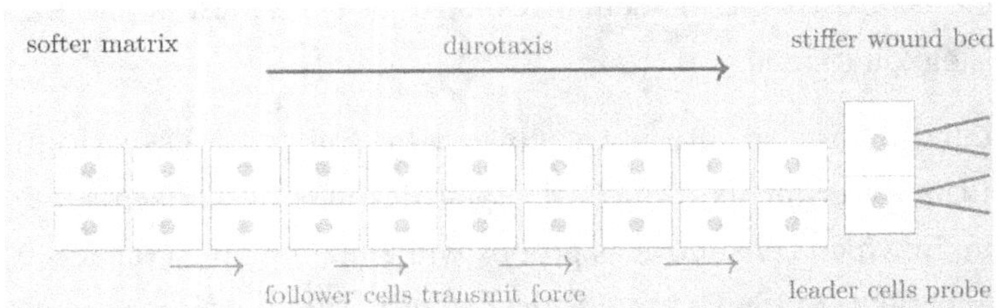


*Collective migration: leaders sense stiffness, followers keep the sheet coherent*

The temptation is to read this diagram as a simple map: “stiff is to the right, so cells go right.” Real wounds are rougher. Stiffness is patchy. Strain is time-varying. The matrix is being remodelled even as cells attempt to use it as a guide. Yet the underlying principle holds: cells can turn mechanical differences into directional decisions.

There is an even subtler cue than stiffness: *strain.* Stiffness is a material property—how much something resists deformation when you push or pull it. Strain is the deformation itself—how stretched or compressed a region is at a given moment. Near a wound, strain fields can be strong and anisotropic because tissue tension has been redistributed and because contracting cells are actively pulling. A cell can, in effect, read the local pattern of deformation and move along it.

This idea is sometimes called *tensotaxis* when migration is guided by gradients of tension or strain rather than stiffness alone. The distinction matters because a region can be soft yet highly strained, or stiff yet relatively unstrained, depending on context. A stretched rubber band is soft but taut; a ceramic tile is stiff but not necessarily under tension. Cells may respond differently to the two situations, and wounds can present both at once. A provisional fibrin network may be compliant but under tension from clot retraction; later collagen may be stiffer and also tensioned by myofibroblasts.

Why would strain guide movement? One reason is that strain changes the architecture of fibres. Pull on a fibrous matrix and the fibres align along the direction of pull. A cell moving through such a matrix tends to align with fibres because adhesions form more readily along fibres than across gaps, and because pulling along aligned fibres transmits force effectively. This is a mechanical version of following the grain in wood. Another reason is that strain changes the forces on adhesions directly. A cell embedded in a pre-strained matrix feels a different baseline load even before it generates its own contraction. That baseline can alter adhesion signalling, cytoskeletal tension, and the stability of protrusions, biasing movement.

This tactile navigation has a paradoxical feature: sometimes the most effective way to "follow" a mechanical cue is to *change* it. Fibroblasts are not passive walkers; they are matrix remodelers. They can locally soften tissue by degrading matrix, locally stiffen it by depositing collagen, and locally align it by pulling. In skin-like matrices, fibroblasts can even reorganise their surroundings in phases, first tugging and rearranging fibres, then changing orientation and moving along the newly prepared tracks. The environment becomes partly self-made. The phrase "the cell shapes the path it takes" risks sounding mystical, but in fibrous materials it is literal: traction forces reconfigure the scaffold, and the reconfigured scaffold guides subsequent traction.

If mechanics can guide so powerfully, why not close every wound as fast as possible? Speed feels like an obvious virtue. Anyone who has watched a child's cut seal over in days and compared it to a stubborn ulcer in old age understands the longing for quick closure. Yet tissues often behave as though closure has an optimal pace, not a maximal one. Part of the reason is immunological—time is needed to clear microbes and debris—but the timing is also regulated mechanically.

Mechanosensitive ion channels provide one window into this regulation. Piezo channels, for example, open in response to membrane tension and allow ions to flow, linking mechanical stress to electrical and biochemical signals. Work in the early 2020s suggested that Piezo can act as a kind of brake on epithelial wound closure in some contexts. When the brake is removed, closure may accelerate, but at a cost: the coordination with inflammatory signalling and reactive oxygen species (ROS) production can be impaired, and the repaired tissue may be less robust. The unsettling implication is that a wound can close *too well* in the short term, sealing before the underlying choreography has ensured a durable barrier.

This is a deeply human trade-off. Think of hurried home repairs: a hole in a wall can be patched quickly with tape and paint, but if water has been leaking behind it, the real damage worsens. Biology faces an analogous problem. Epithelial integrity matters, but so does the quality of the seal and the state of the tissue beneath. Mechanical sensors such as Piezo are well positioned to arbitrate that trade-off because they respond not to abstract instructions but to physical strain at the site of repair. A tissue under high tension may need slower, more coordinated closure to avoid tearing or misalignment; a tissue under low tension may tolerate faster sealing.

Clinical practice already embodies this wisdom, even if it rarely uses mechanobiological language. A surgeon does not simply close a wound as tightly as possible; they aim for approximation without strangulation, leaving enough perfusion for survival of the edges. Burn care involves careful management of tension and movement because early, aggressive contraction can produce disabling contractures. Chronic wound management often begins with controlling shear and pressure—reducing the mechanical insults that keep tissues in a perpetual state of damage and confused signalling. In each case, the aim is to make the mechanical landscape legible and stable enough for cells to navigate and coordinate.

The tactile world of a wound is therefore not a single arrow pointing inward. It is a shifting field of stiffness, strain, fibre alignment, and boundary forces. Cells probe it continually through adhesion dynamics and cytoskeletal tension. As they move, they leave mechanical footprints: aligned fibres, redistributed stresses, locally stiffened tracks. In a successful repair, those footprints converge toward a coherent architecture—one that allows forces to be borne again without constant emergency signalling.

A healed wound is often judged by how it looks, but a cell would judge it by how it feels: whether the matrix resists in the right way when tugged, whether tensions are balanced rather than lopsided, whether the surface can take everyday strain without reopening. Long after redness fades, the tissue still carries a mechanical memory of the breach, written in the direction of fibres and the quiet baseline tension that returns when the gap is no longer a place to fall into.

Chapter 8

# Rogue Mechanics: Cancer, Fibrosis, and Disease

*We often imagine disease as a chemical failure, but the body's collapse is frequently a physical one. From the runaway stiffening of scarred organs to the hydraulic pressures that harden arteries, this chapter exposes how cells misinterpret mechanical cues to build their own prisons.*

## 8.1 The Scar That Never Fades

A cut on the skin is a small drama with a satisfying ending. Blood clots, immune cells arrive, a temporary scaffold forms, and fibroblasts (the body's chief builders of connective tissue) lay down collagen to stitch the gap. Over time the scar softens, reorganizes, and often fades into the background of ordinary life. Fibrosis is what happens when that ending never comes. The same machinery that closes a wound keeps running, not because the body is careless, but because it is listening to a signal that is hard to ignore: the tissue has become mechanically *wrong*— too stiff, too tense, too resistant to deformation. In fibrosis, mechanics does not merely reflect disease.

Mechanics helps *produce* it.

To understand why, it helps to picture connective tissue not as inert packing material but as a dynamic composite. The extracellular matrix (ECM) is a meshwork of proteins and sugars that cells both inhabit and continually remodel. Collagen fibers are the steel rebar; proteoglycans and glycosaminoglycans are the water-binding gel; elastin provides spring. In healthy organs, this mesh is tuned: firm enough to support shape, compliant enough to accommodate movement, breathing, pulsation, and growth. When the ECM stiffens, it changes how forces spread through tissue, and it changes what cells think they are sitting on. The key twist is that fibroblasts are not blind to stiffness. They are, in a very literal sense, mechanosensors.

The wound-healing program includes a particular player, the myofibroblast, which is halfway between a fibroblast and a smooth muscle cell. Myofibroblasts are built for pulling. They express bundles of actin and myosin that generate contractile force, and they anchor these bundles to the matrix via adhesion complexes rich in integrins, focal adhesion kinase (FAK), and many other proteins that translate tugging into biochemical signals. In a normal wound, myofibroblasts are temporary. They cinch the wound closed and lay down collagen, then they disappear, often by programmed cell death, once the job is done. Fibrosis is, in part, a failure of that disappearance.

The failure is not only chemical; it is also physical. Imagine standing on a trampoline versus standing on a concrete floor. On the trampoline, your body's small shifts sink into the surface; the surface gives way and absorbs your weight. On concrete, it does not. Cells experience an analogous difference. A fibroblast constantly pulls on its surroundings through integrins that bind to matrix proteins. If the matrix is soft, that pulling mainly deforms the matrix; the cell cannot build much internal tension. If the matrix is stiff, the matrix resists, and the cell's own cytoskeleton becomes tensed like a tightened rope. The cell interprets that internal tension as information: a message that it is in a rigid environment where extra force and extra reinforcement are required. In healthy repair, this is adaptive. In fibrosis, it is a trap.

A particularly vivid example comes from the lung, an organ designed to be exquisitely deformable. With every breath, the lung stretches and recoils. Now consider what happens after repeated injury: inhaled toxins, infection, autoimmune inflammation, or unknown triggers as in idiopathic pulmonary fibrosis. Collagen accumulates, the tissue stiffens, and a breath becomes not only harder for the person but also mechanically different for the cells. Each inflation stretches a stiffer scaffold, and that stretch is transmitted more efficiently to cells. A mechanobiological feedback loop begins to form: stiffening makes stretch signals stronger; stronger stretch signals promote further stiffening.

The central idea can be drawn as a cycle: a stiff matrix activates myofibroblasts; activated myofibroblasts deposit and remodel collagen; the matrix becomes stiffer; the stiffer matrix further activates myofibroblasts. The loop is powerful because it is self-validating. Once stiffness crosses a threshold, the tissue can begin to behave as though it is still wounded even when the original insult has passed.

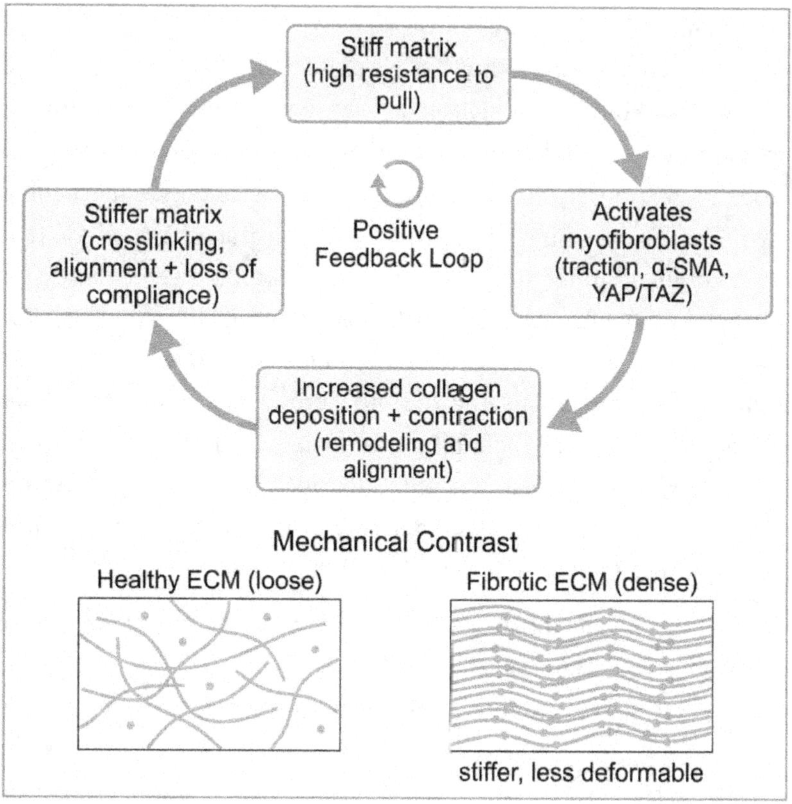


How does a cell "read" stiffness? Not with nerves, not with eyes, but with a molecular handshake and a tug. Integrins span the cell membrane: one end grips the extracellular matrix, the other attaches to the actin cytoskeleton inside the cell via large multiprotein assemblies called focal adhesions. When a cell contracts its actomyosin network, it pulls on the integrin-matrix connection. If the matrix yields, the integrin complexes do not experience much load; they stay small and transient. If the matrix resists, the load stabilizes and enlarges these adhesions, recruiting more signaling proteins and increasing the cell's ability to generate force. The cell turns a mechanical test into a biochemical decision.

Two families of pathways illustrate why that decision can become sticky. One is the Rho/ROCK axis (Rho GTPases and Rho-associated kinases), which increases contractility by regulating myosin activity and actin organization. The other involves YAP and TAZ, transcriptional co-activators that act like mechanical switches. Under high tension and stiff substrates, YAP/TAZ tend to accumulate in the nucleus, where they promote gene-expression programs associated with growth, matrix production, and survival. Under low tension and softer environments, they are more often kept out of the nucleus and the pro-fibrotic program quiets down. Work in the 2010s, including studies led by Fei Liu and David Lagares (2015), helped crystallize the idea that YAP/TAZ are not merely bystanders in fibrosis but part of the circuitry that couples stiff matrices to activated fibroblasts.

Even if one avoids the details of protein names, the logic is intuitive. A builder who steps onto a sagging floor will spread weight and tread carefully; a builder who steps onto solid concrete can brace, push, and lift. A stiffened tissue lets fibroblasts brace and pull harder, which makes them behave as if a stronger repair response is required.

The chemical side of fibrosis often centers on transforming growth factor beta (TGF-$\beta$), a potent signal that encourages fibroblasts to become myofibroblasts and to produce extracellular matrix. The mechanical twist is that TGF-$\beta$ is frequently secreted in a latent form, bound up in a complex that keeps it inactive. It is, in effect, packaged in a way that requires unwrapping. Some of that unwrapping is enzymatic, but some is mechanical. Integrins can bind to the latent complex and, by pulling, help activate TGF-$\beta$ locally. Classic work in the late 1990s showed how an integrin called $\alpha_v\beta_6$ can bind to the latency-associated components of TGF-$\beta_1$ and activate it, providing a mechanism for spatially restricted control. Two decades later, experiments in fibrotic lung tissue and models of stretch (including work by Froese and Shimbori in 2016) showed how mechanical stretch can increase TGF-$\beta$ signaling, linking the physical act of stretching tissue to the biochemical decision to make more scar.

This coupling between force and growth factor matters because it turns mechanical load into a self-reinforcing signal. As the lung becomes stiffer, the same breathing motion can deliver greater effective mechanical stress to cells and latent complexes. In other words, the organ's ordinary motion begins to serve as a daily reminder, delivered in the language of force, that the wound is "still there."

Fibrosis is not limited to the lung. In the liver, chronic injury from hepatitis viruses, alcohol, or metabolic disease activates hepatic stellate cells, which can transform into myofibroblast-like cells and deposit collagen in the space between liver cells and blood vessels. In the kidney, persistent inflammation or hypertension can lead to interstitial fibrosis that gradually erodes filtration. In the heart, after a myocardial infarction, scar is necessary to prevent rupture, but excessive or persistent scarring can stiffen the wall and impair filling. In each case, the organ fails in a manner that reflects its mechanical design: lungs lose compliance, hearts lose diastolic flexibility, kidneys lose the delicate architecture needed for filtration, livers lose the soft permeability needed for blood flow.

A concrete example brings the mechanobiology into focus: mechanical ventilation in intensive care. Positive-pressure ventilation can be life-saving, but it can also injure lungs when pressures and volumes are high, a phenomenon recognized over decades and now described as ventilator-induced lung injury. The mechanobiological lesson is uncomfortable: stretching tissue is not a neutral act. Cells interpret stretch, and in a fibrotic or inflamed context that interpretation can tilt toward further damage and further repair. Modern "lung-protective" ventilation strategies arose from clinical trials and a growing recognition that the lung's mechanics can be pushed into pathological regimes. The story illustrates a general principle: when tissue is already primed by inflammation and stiffness, mechanical forces that would otherwise be tolerated can become instructive signals that amplify pathology.

Historically, physicians described cirrhosis, pulmonary fibrosis, and "hardening" diseases long before they knew anything about integrins or YAP/TAZ. The very language of older medicine often used texture as diagnosis. René Laennec, the French physician who introduced the stethoscope in the early nineteenth century, also contributed to pathological anatomy, describing organs by their feel and appearance at autopsy: granular livers, indurated lungs. Those tactile observations were not merely poetic. They were direct encounters with altered tissue mechanics. Mechanobiology does not replace those observations; it explains how an organ can be pushed into a new material state and then held there by cellular feedback.

The persistence of fibrosis has another mechanical ingredient: collagen is not just deposited; it is organized and chemically strengthened. Enzymes such as lysyl oxidases help crosslink collagen fibers, increasing stiffness and resistance to degradation. Myofibroblasts also align fibers by pulling on them, turning a random mesh into something more like a rope bundle. Aligned fibers can guide cell migration and transmit force over longer distances, allowing mechanical signals to propagate. A local patch of stiff scar can therefore influence neighboring tissue, not only by releasing soluble factors but by changing the stress landscape. Cells several millimeters away can find themselves in a mechanically altered neighborhood.

One reason the fibrotic cycle is so hard to break is that stiffness is a durable memory. Chemical signals can fade quickly. Cytokines diffuse, get degraded, or are shut off. Crosslinked collagen persists. A tissue that has been remodeled into a dense, aligned matrix can continue to present itself to cells as a rigid substrate even after inflammation subsides. This is one sense in which scars "remember" past injuries. The memory is encoded in the material itself.

There is also a subtler time dimension. Living tissues are not purely elastic like a spring; they are viscoelastic, meaning they behave partly like solids and partly like fluids. Under a brief force, tissue may snap back; under a prolonged force, it may slowly creep and remodel. Cells can distinguish these regimes. A matrix that relaxes stress quickly can feel different from one that holds tension. Fibrotic tissues often show altered viscoelastic properties: increased stiffness, reduced ability to dissipate stress, and changes in how forces are shared between the cellular and extracellular components. For a myofibroblast, this can mean that every contraction meets a more stubborn resistance, reinforcing the intracellular tension that keeps mechanosensitive pathways turned on.

It would be comforting to imagine that fibrosis is simply too much collagen, as though a single knob were turned too far. The reality is closer to a shift in a whole mechanical ecosystem. Fibroblasts change their cytoskeleton to generate more force. They increase adhesion strength to hold onto the matrix. They secrete more matrix, and they secrete enzymes that crosslink it. They change the balance of matrix-degrading enzymes and their inhibitors. Immune cells contribute signals that keep fibroblasts activated. Blood vessels may be compressed or remodeled, changing oxygen and nutrient delivery, which can further influence cell behavior. Each element can be described biochemically, but the mechanical picture adds a unifying theme: force and stiffness are not incidental side effects; they are shared currencies that many pathways can deposit into and withdraw from.

Therapeutically, this framing suggests why fibrosis is so stubborn. If one blocks a single cytokine, the mechanical environment may still keep cells in an activated state. If one reduces inflammation, the crosslinked matrix can remain as a rigid cue. If one softens the matrix but cells still generate high contractile force, they may re-stiffen it. The most promising interventions may be those that interrupt the feedback at multiple points, or those that target the mechanotransduction machinery that links stiffness to gene expression. The idea has motivated attempts to inhibit integrin-mediated activation of TGF-$\beta$, to modulate Rho/ROCK signaling and cellular contractility, and to interfere with YAP/TAZ-driven transcriptional programs. Even when such approaches do not become cures, they sharpen the question that fibrosis poses: how can an organ be persuaded to stop repairing itself?

There is a human cost to getting this wrong. Pulmonary fibrosis can turn a short walk into breathlessness and an evening conversation into exhaustion. Liver fibrosis can progress silently until complications appear, when the architecture is already altered beyond easy repair. Chronic kidney fibrosis can narrow a life around dialysis schedules. These conditions feel slow, almost geological, which is fitting: fibrosis is biology that has started behaving like materials science. The tragedy is that the body is not trying to sabotage itself. It is trying to make a stable patch. The stable patch becomes a stable prison.

The scar that never fades is not merely the presence of collagen; it is the presence of a particular *mechanical logic*. A stiff matrix tells fibroblasts to become myofibroblasts. Myofibroblasts pull harder and build more matrix. The matrix stiffens further and holds onto that stiffness through alignment and crosslinking. Normal motion—breathing, blood flow, organ stretch—becomes a periodic input that can keep mechanosensitive signals active. The loop is persuasive because each step makes the next more likely, as though the tissue has learned a single lesson too well: when in doubt, reinforce.

A healthy scar is a compromise between strength and flexibility. Fibrosis is what happens when strength keeps winning that argument, one collagen fiber at a time, until the organ can no longer do the soft, subtle work it was built for.

## 8.2 The Engine of Malignancy

A tumor can be felt before it is understood. The classic breast lump is, in a blunt sense, a mechanical diagnosis: something inside the body has become *hard*. Long before genes and microscopes entered the story, physicians used the ancient word *carcinos* for cancers that seemed to grip tissue like a crab. The metaphor is visual, but the experience is tactile. Malignancy announces itself as a change in material.

That hardness is not a cosmetic detail. It shapes what cancer cells do, what nearby normal cells are pressured into doing, and what therapies can reach. Cancer is often told as a tale of mutations and runaway cell division. Those mutations are real, but they act inside a physical world. Every new cell must make room for itself. Every migrating cell must squeeze through pores, detach and reattach, and survive forces that would rip most cells apart. Even the biochemical signals that drive growth are influenced by whether a cell sits in something soft like brain tissue or fibrous like the breast. The tumor is not only a genetic renegade; it is a mechanical engine that converts growth into stress, and stress back into growth.

The most obvious mechanical feature of many solid tumors is their abnormal extracellular matrix (ECM). Healthy tissue ECM is a supportive mesh that is renewed and reorganized without becoming a cage. In many cancers, especially those with a strong stromal response, the ECM becomes dense, aligned, and crosslinked. Pathologists sometimes describe the surrounding tissue as *desmoplastic*: a scarlike reaction with abundant collagen and activated fibroblasts.

The result is often measurable stiffness, and that stiffness changes the rules of cell behavior.

The key mechanobiological move is familiar from wound repair: cells pull on their surroundings through integrins, and what they feel determines what they become. Integrins are the cell's gripping tools, linking ECM proteins outside to actin cables inside. When the ECM is stiff, pulling does not simply deform the environment; it loads the cell's internal skeleton, stabilizes focal adhesions, and triggers a cascade of signaling. In 2005, Matthew Paszek and colleagues captured this with a striking concept: *tensional homeostasis*. Epithelial tissues are normally organized around a balance of forces that preserves polarity and architecture. When the ECM stiffens, that balance shifts. Cells generate more contractile force through the Rho/ROCK pathway, adhesions enlarge, and growth signals such as ERK become easier to sustain. In a three-dimensional gland, the consequence can be architectural: normal hollow structures that should form orderly lumens instead become disorganized and filled in, a mechanical push away from healthy form.

This is one reason tumors and fibrosis so often resemble each other in their physical logic. In fibrotic disease, stiffness recruits and locks in myofibroblasts that deposit more collagen. In cancer, tumor cells live inside, and help construct, a stiffened niche that rewards aggressive behavior. Fibroblasts in the tumor microenvironment can become *cancer-associated fibroblasts* (CAFs), which behave much like myofibroblasts: contractile, matrix-producing, and capable of reorganizing collagen into aligned "tracks." Those tracks are not neutral; they can function as highways that guide invasion.

The word *invasion* tempts a purely biochemical picture: enzymes dissolve barriers and cells crawl through. But the act is also mechanical. A basement membrane is a thin, tough sheet of ECM that normally separates epithelia from underlying connective tissue. For a carcinoma cell to metastasize, it must cross that boundary, traverse the stroma, enter blood or lymphatic vessels, survive shear stresses in circulation, exit elsewhere, and then establish a colony in a new tissue. At every stage, force is part of the price of admission.

A concrete image helps. Cancer cells extend actin-rich protrusions called invadopodia, which concentrate enzymes that digest ECM. Yet invadopodia are not just chemical drill bits. They are also force-generating structures that push and pull, probing for weak spots, widening pores, and deforming the matrix. A cell does not slip through a dense collagen network like a ghost; it shoves, twists, and squeezes.

The nucleus is often the limiting object in that squeeze. It is large, relatively stiff, and packed with chromatin. A migrating cancer cell can flatten its cytoplasm dramatically, but the nucleus must pass through whatever opening exists. When pores are smaller than the nucleus, the cell can either enlarge the pore (by proteolysis and mechanical tearing) or deform the nucleus. Deforming the nucleus is risky. It can rupture the nuclear envelope, expose DNA to damage, and create a selection pressure for cells that can survive such insults. Mechanical constraints can therefore shape evolution inside the tumor: the microenvironment acts as a filter that favors cells with particular cytoskeletal and nuclear properties.

This makes a profound point about metastasis. It is not only a matter of "deciding" to migrate in genetic terms. It is also a matter of being mechanically equipped to do so.

Stiffness is not the only mechanical feature of tumors. Tumors also generate *stress*. As a mass grows inside a confined organ, it pushes on everything around it. This is solid stress: compressive forces stored in the tissue, like tension in a stretched rubber band or compression in a packed sponge. At the same time, many tumors have leaky blood vessels and impaired lymphatic drainage, leading to elevated interstitial fluid pressure. These two pressures—solid and fluid—are related but not identical. A tumor can be mechanically oppressive both as a solid object and as a pressurized wet sponge.

The consequences are practical and cruel. Solid stress can compress blood vessels, reducing perfusion. Poor perfusion means low oxygen (hypoxia) and limited delivery of drugs. Hypoxia can select for more aggressive phenotypes and can make radiotherapy less effective, since oxygen influences radiation-induced DNA damage. Elevated interstitial fluid pressure can reduce convective transport into the tumor, further limiting how therapeutics penetrate. Triantafyllos Stylianopoulos and others have argued that the physics of stress and pressure is not peripheral to cancer treatment; it can be a major reason tumors resist therapy even when drugs are potent in a dish.

These mechanical features also feed back into cell signaling. Compressive stress can alter cytoskeletal organization and influence proliferation and apoptosis resistance. Cells under pressure can activate pathways that help them survive being crowded and squeezed. Crowding itself changes how cells sense contact with neighbors, how they polarize, and whether they detach. Tumor mechanics is therefore not just a problem for clinicians; it is a continuous input into the tumor's internal logic.

The historical anecdote that captures the tactile side of all this comes from a surgeon's hands. In the late nineteenth and early twentieth centuries, as modern surgical oncology developed, surgeons learned to judge margins not only visually but by palpation. "Hard" tissue suggested invasion; softer planes suggested separations that could be respected. This was not superstition. It was an empirical recognition that cancers often remodel their surroundings into a different material, and that the transition from healthy to malignant is not only cellular but architectural. Even today, imaging modalities such as elastography exploit this same fact by mapping tissue stiffness to help detect tumors. The body's mechanical map can be diagnostic because malignancy rewrites it.

One might still ask a skeptical question: is stiffness a driver of cancer, or merely a consequence? The honest answer is that it can be both, depending on context, and the two roles can reinforce each other. Some of the strongest evidence for stiffness as a driver comes from experiments that manipulate the matrix itself. In 2009, Kandice Levental and colleagues showed that increasing collagen crosslinking in breast tissue can stiffen the ECM and promote malignant progression in models, enhancing integrin signaling and invasion. This kind of result matters because it shifts stiffness from symptom to cause: change the mechanics, and you change the biology.

Crosslinking is especially potent because it hardens the matrix without necessarily adding more material. It is the difference between a loose rope net and the same net soaked in glue and allowed to cure. Collagen crosslinking enzymes, including lysyl oxidase (LOX) family members, can increase fiber stiffness and stabilize aligned architecture. Cells respond by building larger focal adhesions and sustaining growth and motility signals. A mechanically hardened niche becomes a niche that rewards the traits we fear in cancer: persistence, invasion, survival under stress.

The mechanical microenvironment also shapes how tumors interact with the immune system. Immune cells must migrate through stroma to reach cancer cells. Dense, aligned collagen can act as a physical barrier, and high interstitial pressure can influence transport of signaling molecules. While immunology has its own rich language of checkpoints and antigen presentation, the physical layout of the battlefield matters. A tumor that has remodeled its ECM can alter who gets in, who gets out, and who can make contact.

Mechanics even complicates the seemingly straightforward notion of "growth." A proliferating cell colony in a dish expands because space is free. In tissues, space is earned. As cells proliferate, they generate compressive forces. Those forces can slow proliferation in some contexts (contact inhibition), but cancer cells may bypass normal inhibition. Yet bypassing inhibition does not remove the physics; it means growth continues into resistance, increasing internal stress. That stress can then encourage invasive escape, as cells seek paths of lower mechanical resistance. In this view, invasion is not only a biochemical program but also a pressure-release strategy.

There is an almost industrial efficiency to the way tumors exploit mechanical feedback. Stiffness and stress promote signals that increase contractility. Increased contractility helps cells remodel and align the matrix. Aligned and crosslinked matrix increases stiffness and provides tracks for migration. Migration and invasion relieve local crowding and open new territories. Meanwhile, stress and pressure distort vasculature, creating hypoxic zones that select for hardier cells and impair therapy delivery. The tumor is not consciously designing this system, but natural selection within the tumor favors cells and stromal partners that thrive inside it.

One of the most unsettling aspects of this story is that many of the same mechanobiological tools are used in normal development and repair. Integrins, FAK, Rho/ROCK, YAP/TAZ, actomyosin contractility—these are not cancer inventions. They are ancient solutions to living in a physical world. Cancer weaponizes them by uncoupling them from restraint. Where a wound-healing response is normally temporary and bounded, the tumor microenvironment can keep supplying the mechanical cues that maintain activation. Where normal epithelial tissues maintain polarity despite fluctuations in force, stiffened environments can push them toward disorganization. Where cells normally migrate in controlled contexts, cancer cells exploit the same machinery to travel without permission.

The picture also helps explain why metastasis is so inefficient and yet so deadly. Most cells that enter circulation die. Many that exit elsewhere fail to colonize. But the few that succeed often share mechanical competencies: the ability to deform without rupturing fatally, the ability to generate traction in unfamiliar matrices, and the ability to survive transient detachment (anoikis resistance). The metastatic cell is not only genetically altered; it is mechanically tough and mechanically opportunistic.

Consider a breast carcinoma cell leaving its primary site. It must cross the basement membrane, which is not merely a chemical barrier but a tensile sheet. It must traverse collagen-rich stroma, where pore sizes may be smaller than its nucleus. It must enter a vessel by squeezing between endothelial cells or exploiting leaky vasculature. Once in the bloodstream, it faces shear stress, collisions with blood cells, and rapid changes in pressure and oxygenation. Arrest in a capillary requires deformation and sometimes active adhesion. Extravasation requires another mechanical negotiation with the vessel wall and surrounding matrix. At no point does "motility" reduce to a single gene. It is a choreography of force.

If all this sounds bleak, it also offers a kind of clarity. Mechanical features can be measured, manipulated, and sometimes normalized. Strategies that aim to "soften" the tumor stroma, reduce solid stress, or alter crosslinking are attempts to change the tumor's physical incentives. By decompressing vessels, one may improve drug delivery. By interrupting integrin-mediated mechanotransduction, one may reduce the ability of cells to interpret stiffness as a growth signal. The challenge is that the same pathways matter in healthy tissues, so therapies must be both targeted and timed.

A final image returns to the feel of a tumor. When a clinician palpates a hard mass, they are touching the end result of countless microscopic tugs: fibroblasts pulling, cancer cells contracting, collagen fibers aligning, enzymes crosslinking, vessels compressing. The lump is a macroscopic summary of mechanobiology. It is tissue that has been remodeled into a material that favors persistence over flexibility. That material does not merely sit there. It pushes back, and in pushing back it teaches cells how to behave.

## 8.3 Pressure Points

The circulatory system is the body's most relentless mechanical device. It does not rest when you sleep, or when you are ill, or when you are afraid. With each heartbeat it pushes fluid through a branching, elastic network whose walls must be strong enough to contain pressure and responsive enough to adapt. When those walls fail, they do so in ways that sound like engineering reports: aneurysm, rupture, stenosis, leakage, stiffening. Cardiovascular disease is full of chemistry, but it is also full of hydraulics.

Hydraulics begins with a simple fact: blood is a fluid, and fluids transmit forces. Pressure is force spread over area, and blood pressure is the push that drives flow. The vessels are not rigid pipes. They are living tubes, lined by endothelial cells that directly touch the blood and backed by smooth muscle cells and connective tissue that give the wall strength. The heart supplies pulsatile pressure, and arteries act as elastic reservoirs that smooth that pulsation into steadier flow downstream. In an idealized picture, blood travels in neat layers, with the fastest layer in the center of the vessel and the slowest near the wall. That layered movement is called *laminar flow*. The real system is messier, especially at branches and curves, but laminar flow is still the baseline the vessel wall expects.

Endothelial cells have a peculiar job. They must be both barrier and sensor: tight enough to keep blood where it belongs, and perceptive enough to detect what the flowing blood is doing. Their most constant mechanical input is *shear stress*, the tangential frictional force created by blood sliding past the vessel wall. Shear stress is not the same as pressure. Pressure pushes outward, trying to inflate the vessel like a balloon. Shear scrapes along the surface, like a current against a riverbank. Endothelial cells can feel both, and they respond differently.

Under steady laminar shear, endothelial cells tend to align in the direction of flow, elongating like reeds in a stream. Their internal cytoskeleton reorganizes, their gene expression shifts, and the vessel adopts a calmer, more anti-inflammatory state. Under disturbed flow, which can include oscillations, turbulence, and flow separation near branches, endothelial cells often become disorganized, less aligned, and more prone to express inflammatory signals. This is one reason atherosclerosis has favorite sites: not randomly distributed, but clustered at arterial bends, bifurcations, and regions where flow is complex. Cholesterol and immune cells matter enormously, yet geometry and shear decide where the wall is most vulnerable.

The link between disturbed flow and vascular disease has been appreciated for a long time, but the story gained a sharper edge in the twentieth century as fluid mechanics matured. A useful historical anecdote begins with a young medical student, William Harvey, in the early seventeenth century. Harvey did not have microscopes or pressure sensors, but he had an experimental instinct and a talent for quantitative reasoning. In 1628 he published *De Motu Cordis*, arguing that the heart acts as a pump and that blood circulates. His reasoning included estimates of how much blood the heart could move, concluding that the body could not possibly be manufacturing and consuming that volume anew. The circulation, in other words, had to be a loop. Harvey's insight was not mechanobiology in the modern sense, but it set the stage for seeing the cardiovascular system as a problem of motion, force, and volume.

That mechanistic view has only deepened. One can treat large arteries as composite materials under cyclic load. The wall's collagen fibers provide tensile strength; elastin provides recoil; smooth muscle cells adjust diameter and stiffness moment to moment. The system works beautifully until it doesn't, and a central failure mode is *hypertension*: chronically elevated arterial pressure. Hypertension increases the outward force on vessel walls. In response, vessels remodel. The wall may thicken, smooth muscle may become more contractile, and the ECM may change in composition and crosslinking. These changes can be adaptive at first, aimed at normalizing wall stress, but over time they often stiffen arteries and further elevate systolic pressure, placing more burden on the heart.

This is a mechanical feedback loop with clinical consequences. Stiffer arteries do not cushion the heart's pulse as effectively. The pressure wave travels faster and reflects more strongly from branch points, raising systolic blood pressure and widening pulse pressure. The left ventricle must push against this increased afterload, and it may hypertrophy. Eventually the heart can fail not because it cannot contract, but because it cannot relax and fill properly, a common form of heart failure in older adults.

The endothelium sits at the center of this, because it interprets flow. Under healthy laminar shear, endothelial cells tend to produce nitric oxide, a molecule that relaxes smooth muscle and discourages platelets from sticking. Under disturbed shear, they can produce less nitric oxide and more adhesion molecules that invite immune cells to attach and crawl into the vessel wall. Over years, this can seed atherosclerotic plaques. The plaque itself changes mechanics again: it alters vessel geometry, further disturbing flow, and it creates regions of stress concentration that influence whether the plaque remains stable or ruptures. Plaque rupture is a mechanical event with biochemical triggers, and when it happens it can abruptly occlude a coronary artery, causing a heart attack.

It is tempting to imagine that the vessel wall is passively battered by flow, but mechanobiology insists on agency. Cells are not just damaged by forces; they *interpret* them. Endothelial cells have mechanosensors embedded in their membranes and junctions: ion channels that open under deformation, adhesion complexes that shift under shear, and glycocalyx components that bend in flow. Mechanosensing is distributed rather than centralized, like a building whose stability depends on many beams rather than a single pillar.

Among the most intriguing candidates for direct mechanosensors are mechanosensitive ion channels, especially the Piezo family. Piezo channels are large membrane proteins that open when the membrane is stretched, allowing ions to flow in and rapidly change cellular signaling. In the 2010s, experiments in mice provided dramatic evidence that Piezo1 and Piezo2 in baroreceptor neurons are crucial for the baroreflex, the rapid feedback loop that stabilizes blood pressure from moment to moment. Baroreceptors are pressure-sensing nerve endings in the carotid sinus and aortic arch; when blood pressure rises, they increase firing, and the brainstem adjusts heart rate and vessel tone to bring pressure down. When this reflex fails, blood pressure becomes dangerously labile, surging and crashing with posture, stress, and routine activity.

The Piezo story is appealing because it seems to offer a clean mechanistic link: pressure stretches the arterial wall, stretch opens Piezo channels in baroreceptor endings, the resulting electrical activity signals the brain, and the baroreflex follows. It is the kind of mechanism that feels like it belongs in a schematic diagram. Yet biology rarely grants a single-master-switch explanation without caveats. There has been thoughtful skepticism in the literature about whether Piezo1/2 alone satisfy every criterion for being *the* baroreceptor mechanosensor, rather than *a* key component in a larger mechanosensory apparatus. Baroreceptor endings are embedded in a complex mechanical environment involving extracellular matrix, connective tissue scaffolds, and potentially multiple channels and receptor systems. The controversy is productive: it keeps attention on the fact that a sensory neuron is not merely a wire but a mechanically coupled structure whose sensitivity depends on how it is tethered and loaded.

Whatever the final accounting, the baroreflex makes a crucial point for mechanobiology: the cardiovascular system does not only endure forces; it uses forces as information. Blood pressure is not just a number. It is an input that is sensed, computed, and acted upon in real time.

There is another place where mechanics and disease intersect in a particularly physical way: the microcirculation, where arteries branch into arterioles and capillaries. Capillaries are so narrow that red blood cells deform to pass through. This deformation is not trivial; it is an everyday mechanical challenge that red blood cells are specially built to meet. In sickle cell disease, mutations in hemoglobin cause red blood cells to become stiff and misshapen under low oxygen. The result is a mechanical catastrophe at the microvascular scale: rigid cells obstruct capillaries, flow slows, tissues become hypoxic, inflammation increases, and pain crises occur. The molecular cause is well known, but the immediate pathophysiology is a problem of material properties and flow resistance. It is one of the clearest demonstrations that small changes in cellular mechanics can cascade into systemic disease.

Flow patterns also influence how vessels grow and remodel. During development, the vascular network is refined in part by flow: vessels that carry more blood tend to enlarge and stabilize, while those that carry less may regress. In adults, chronic changes in flow can induce remodeling. If an artery is partially blocked, downstream tissues may recruit collateral vessels, and shear stress changes can guide that adaptation. Conversely, abnormal flow can contribute to pathological remodeling in pulmonary hypertension, where the vessels in the lungs constrict and thicken, raising resistance and forcing the right ventricle to work harder. Again, pressure begets remodeling, remodeling alters mechanics, and altered mechanics feeds back into pressure.

The interplay between pressure and vessel health becomes especially poignant when considering aneurysms. An aneurysm is a localized dilation of an artery, often in the aorta. As the diameter increases, the wall tension rises, according to simple mechanical principles: a larger radius under the same pressure produces greater circumferential stress. This creates a dangerous feedback loop: dilation increases wall stress, which can promote further dilation and weakening, increasing the risk of rupture. Surgical decisions about aneurysm repair are, in part, decisions about mechanical thresholds: at what diameter does the risk of rupture exceed the risk of intervention? The body is making a material failure more likely, and clinicians must intervene before the physics wins.

The theme of hydraulic stress also appears in cancer, though in a different register. Tumors often elevate interstitial fluid pressure because their vessels are leaky and their lymphatic drainage is poor. This pressure can flatten nearby vessels and reduce drug delivery, but it also creates gradients that shape how molecules move. Large molecules, including many antibody-based therapies, rely on a mix of diffusion and convection to reach their targets. When interstitial pressure is high and fairly uniform inside the tumor, the driving force for inward convection from blood vessels is reduced. The tumor becomes, hydraulically, a poorly irrigated sponge. Treatment then becomes not only a matter of finding the right molecular target but also of overcoming a pressure-driven transport barrier.

Efforts to quantify tumor mechanics have pushed researchers to define measurable quantities like solid stress and stored elastic energy as biomarkers of a tumor's mechanical pathology. The aspiration is practical: if a tumor's stress state predicts vessel compression and poor perfusion, then measuring that stress could help guide therapies that aim to decompress the tumor or normalize its stroma. The underlying point is broader than oncology. In many diseases, measuring mechanics can reveal what is otherwise invisible: the distribution of forces that cells are responding to.

A person living with hypertension rarely thinks of themselves as hosting a hydraulic machine. They feel stress in a psychological sense, or fatigue, or nothing at all until damage accumulates. Yet the vessels are constantly doing arithmetic: balancing pressure against wall strength, shear against endothelial alignment, pulsation against elastic recoil. Mechanobiology adds that the vessel wall is also writing a long-term memory of these forces. Chronic high pressure encourages collagen deposition and crosslinking, which stiffen the wall. Chronic disturbed shear encourages inflammation and plaque formation. These are not abstract risks; they are cellular responses to mechanical information.

The intimacy of this becomes obvious in small actions. Standing up quickly shifts blood downward. Without the baroreflex, the brain's perfusion would drop, and consciousness would flicker. Running increases cardiac output, raising flow and shear, and the endothelium responds by dilating vessels to deliver oxygen. A narrowed coronary artery changes local flow patterns, and plaques grow preferentially at sites where the wall is mechanically and biologically primed. The body lives on the edge of hydraulic instability, kept safe by layers of mechanosensing and remodeling.

When those layers fail, the failure can be silent for decades and then sudden. A plaque ruptures at a mechanically stressed cap, and a clot forms within minutes. An aneurysm reaches a critical size, and the wall tears. A stiffened arterial tree raises the heart's workload until it cannot keep up. These events feel dramatic because they are, quite literally, failures of structure under load.

Vessels, like all living tissues, are not designed to be unchanging. They are designed to be adaptable, and adaptation is an inherently mechanical process: cells sense forces, change gene expression, secrete matrix, adjust tone, and reconfigure shape. The same adaptability that lets a pregnant uterus receive vastly increased blood flow, or lets an athlete's heart and vessels remodel under training, also means that chronic pathological forces can sculpt the system into a harmful form. The vascular wall is an intelligent material, and it will do what intelligent materials do: respond to the loads it bears, even when those loads are slowly teaching it the wrong lesson.

## Chapter 9

# Probing the Invisible: Tools to Measure and Control Force

*For centuries, biology was a purely visual discipline, restricted to describing the shapes and structures visible through a glass lens. We have since learned to 'touch' the microscopic world, translating the invisible language of piconewtons into tangible maps of tension and stiffness. By turning soft gels into sensors and light beams into mechanical switches, we can now not only observe cellular forces but actively command them.*

## 9.1 Tracking the Invisible Grip

A crawling cell looks, at first glance, like a soft bag of fluid inching across glass. Under the microscope it can seem almost weightless: a thin veil spreading, a rounded tail retracting, the whole body reorganizing as if it were made of smoke. Yet that motion is powered by something stubbornly physical. The cell must push and pull against its surroundings, and it can only move if it can find purchase. In other words: it does not float; it grips.

The difficulty is that the grip is largely invisible. If a person climbs a wall, the wall records the event in scuffed paint, chalk marks, maybe even a new crack. A cell leaves almost nothing behind. When it tugs on the extracellular matrix (the meshwork of proteins and sugars that fills tissues), the matrix deforms by nanometres to micrometres—real distances, but small enough that the eye cannot easily translate them into *force*. And force is what we want, because force tells us not only *that* a cell is active, but *how* it is acting: where it is pulling, in what direction, with what intensity, and whether those pulls are coordinated like a team of rowers or chaotic like a crowd pushing at a door.

Traction Force Microscopy (TFM) was invented to make that hidden grip legible. The basic idea is deceptively simple: give the cell a surface that behaves like a soft trampoline, sprinkle that trampoline with tiny markers, and watch what happens when the cell walks across it. The markers are usually fluorescent beads—microscopic dots that glow under the right light. They are embedded in, or attached to, a soft elastic gel (often a polyacrylamide gel, sometimes silicone-based materials), and the gel's stiffness is chosen to mimic something like soft tissue rather than glass. The cell adheres to proteins on the gel surface (collagen, fibronectin, and so on), spreads, and starts to contract its internal machinery. As it pulls, the gel deforms; as the gel deforms, the beads move. If the beads move, the cell must have exerted traction forces.

That part is intuitive, almost childlike. If you stand on a trampoline, the woven surface sags beneath your feet. If you walk, your footfalls leave shifting patterns of tension that ripple outward. Now imagine you have painted a neat grid of dots on the trampoline. After you take a step, the grid is no longer perfectly regular: dots near your foot have shifted, some closer together, others farther apart. Those dot displacements are footprints made of geometry. TFM turns those geometric footprints into a map of force.

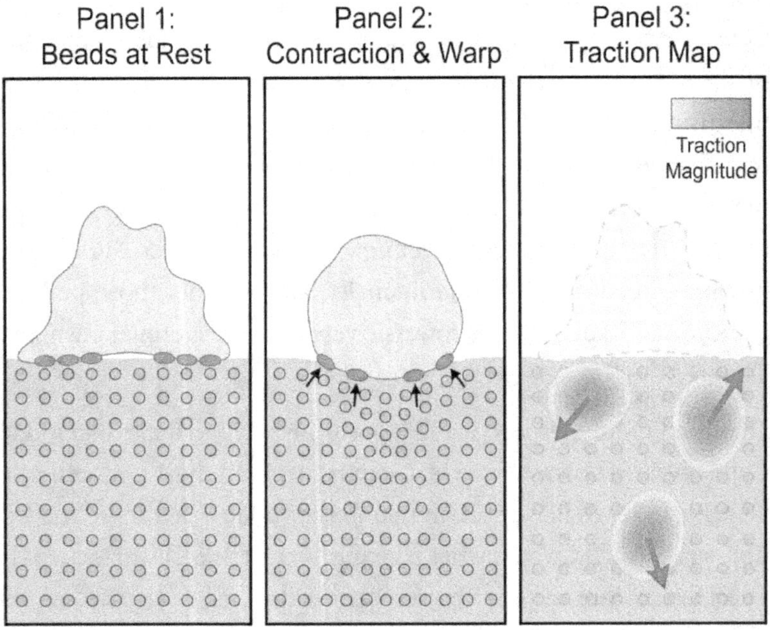


To turn bead shifts into a traction map, TFM needs a translation step: an elastic model of the gel. The gel is treated, in the simplest case, as a uniform elastic material. Elastic means it behaves like a spring: deform it and it pushes back; remove the force and it relaxes. The stiffness of that spring-like behavior is captured by a parameter called *Young's modulus*. When you buy running shoes, you can feel the difference between a soft foam and a firm one; Young's modulus is the number that quantifies that intuition. Brain-like gels can be hundreds of times softer than muscle-like gels; glass is vastly stiffer than either. TFM works best when the substrate is soft enough that cellular pulls produce measurable displacements, but not so soft that the gel behaves like a thick liquid or tears under the cell.

The beads are the reporter pixels. A modern microscope can locate their centres with surprising precision, often far smaller than the bead's diameter, by fitting the brightness profile. First an image is taken while the cell is actively pulling. Then a reference image is taken in a "relaxed" state. The usual trick is to detach the cell (using a chemical that disrupts adhesion, or by killing the cell quickly) so the gel relaxes, and the beads return close to their original positions. Subtract the two images, and every bead displacement becomes a tiny arrow: a direction and a distance. The field of displacements is a footprint field.

But the aim is not a footprint map; it is a traction map: the distribution of stresses the cell exerts at its interface with the substrate. A *stress* is force per unit area, and thinking in stress is often more revealing than thinking in raw force. A cell may apply the same total force in two very different ways: spread thinly like a gentle hand on a table, or concentrated at a few points like fingertips. Both could sum to the same net force, yet their biological consequences differ. Concentrated stresses can tear molecular bonds, remodel matrix fibres, or trigger mechanosensitive signalling complexes; diffuse stresses might merely steady the cell.

Here comes the mathematical sting in the tail. The gel does not tell you forces directly; it tells you displacements. Many different patterns of traction forces could, in principle, produce very similar displacement patterns once the gel spreads the deformation. The problem resembles hearing a muffled sound through a wall and trying to infer the exact motion of the speaker's lips. In technical language, traction reconstruction is an *inverse problem*, and it is often *ill-posed*: small measurement noise in bead positions can lead to large errors in inferred forces unless one adds constraints. Those constraints are imposed through *regularization*, a controlled way of saying, "prefer the simplest force pattern that still explains the data." Regularization is not cheating; it is acknowledging that measurement has limits and that biological force fields do not fluctuate wildly from pixel to pixel without physical reason.

One of the influential advances was to make this inverse problem tractable and fast by moving into Fourier space, an approach often called Fourier-transform traction cytometry. In Fourier space, complicated spatial patterns are re-expressed as sums of waves, and the relationship between displacement and traction becomes algebraically simpler. This is less mystical than it sounds. A piano chord can be decomposed into pure tones; likewise, a complicated map of bead shifts can be decomposed into spatial frequencies. The computation becomes efficient enough that traction maps can be produced routinely, and the method can be paired with time-lapse imaging to watch forces change as cells migrate, divide, or respond to drugs.

The story of why anyone bothered to do this is tied to a simple observation: migrating cells do not move like miniature cars with wheels. Their motion comes from internal contraction, transmitted through adhesion sites into the environment. In the late twentieth century, cell biologists could already see some of the players under the microscope: actin filaments forming bundles, myosin motors generating tension, and focal adhesions acting as molecular rivets. Yet the actual distribution of force remained speculative. Did the cell pull mostly at the back, like a person dragging themselves forward by their elbows? Did it pull mostly at the front, like a climber hauling on a grip? Or did it work more like a caterpillar, with alternating grip and release?

A landmark set of traction maps by Micah Dembo and Yu-Li Wang (1999) on migrating fibroblasts helped settle the argument in a particularly vivid way. Fibroblasts are the workhorse cells of connective tissue and wound repair, and in a dish they crawl with a purposeful, elongated shape. Their traction maps showed strong, localized forces concentrated near the leading edge and along adhesion-rich regions rather than a simple rearward "pull." The cell did not merely tug itself from behind; it generated a complex pattern of pushes and pulls that could, depending on context, look like a coordinated set of anchors and winches. Importantly, the numbers were not cosmic: traction stresses were often in the range of tens to hundreds of pascals (a pascal is one newton per square metre), but concentrated over tiny areas. Multiply stress by area and the resulting forces at individual adhesion sites land in the nanonewton range. A nanonewton is the weight of a small bacterium under Earth's gravity. That is the scale of a cell's handshake with its world: not zero, not vague, but measurable and consequential.

Once you can measure forces, you can ask sharper biological questions. Consider wound healing. Fibroblasts move into a wound bed, lay down new matrix, and contract the tissue, helping close the gap. Clinicians know that contraction can be too little (chronic wounds) or too much (fibrosis and scarring). TFM gives a way to quantify, under controlled conditions, how strongly fibroblasts pull as they differentiate into more contractile forms, and how drugs or genetic changes modulate that pull. It also exposes a subtlety: a more contractile cell is not merely one that generates more force; it may redistribute force differently, shifting from broad, exploratory tractions to focused, sustained ones that remodel matrix.

TFM also illuminates the conversation between a cell and the stiffness of its environment. Place the same type of cell on a very soft gel and on a stiffer gel, and it often spreads more on the stiff one, builds larger adhesions, and generates greater traction. This has become a central theme in mechanobiology: cells adjust their behaviour based on mechanical feedback. The traction map is, in effect, a readout of a decision the cell has made about how hard to "try." But it is not only the cell making a decision. The substrate stiffness alters what the same internal contractile machinery can achieve. Pull on a marshmallow and it yields; pull on a rubber eraser and it resists. In tissues, that difference can separate normal development from disease. Many tumours, for example, are stiffer than surrounding tissue, and that stiffening can feed back onto cell behaviour, encouraging invasion and abnormal growth. TFM does not diagnose cancer by itself, but it provides a mechanistic bridge between the physical property of tissue and the biological behaviour of cells.

In practice, making TFM work requires care that is easy to underestimate. The gel must be thick enough that the cell's pulls do not "feel" the rigid glass beneath it, yet thin enough to image beads clearly. The surface chemistry must allow adhesion proteins to attach without altering the gel's elasticity. Beads must be dense enough to give spatial resolution, but not so dense that they blur together. Even the assumption of simple linear elasticity can fail if the gel is strained too much or if it behaves viscoelastically (partly elastic, partly time-dependent like a slow spring). Cells do not pull in only one direction either: some traction has vertical components, tugging up or pushing down. Standard TFM often focuses on in-plane forces because they are easier to measure from bead shifts in a single imaging plane, but more sophisticated variants reconstruct three-dimensional tractions using volumetric bead imaging.

Then there is the conceptual trap: a traction map is not a direct map of "how strong the cell is" in the way a bathroom scale measures a person. Strength here is distributed, dynamic, and strategic. Two cells can have similar peak traction stresses yet behave differently because one concentrates those stresses at a few long-lived adhesions, while the other flickers through many short-lived ones. This is why mechanobiologists often compute additional summaries from traction fields: not only the maximum stress, but the total strain energy stored in the substrate, or the net contractile moment (a measure of how much the cell acts like a contracting dipole, pulling inward from multiple sides). These are different lenses on the same phenomenon, and they help connect force maps to biological outcomes like migration speed, adhesion turnover, or changes in gene expression.

It is tempting to imagine that the main virtue of TFM is simply that it produces beautiful, colourful maps. But the deeper virtue is that it transforms a qualitative story—"cells pull on their surroundings"—into a quantitative language that can be argued over. When a drug reduces migration, is it because the cell's engine (actomyosin contraction) is weaker, or because its clutch (adhesion) slips? TFM can often distinguish these: a cell with a strong engine but poor clutch may generate internal tension without transmitting it, leading to low traction; a cell with strong traction but slow movement may be over-adhered, stuck like shoes in mud. When a mutation changes cell shape, is it altering force generation or force distribution? Again, the traction map carries the answer.

There is also a quiet philosophical shift embedded in TFM. Biology has long been comfortable with signals that are chemical: concentrations, gradients, binding affinities. Forces demand a different intuition. A chemical gradient can be present without doing anything unless a receptor reads it. A force, by contrast, *is* an action: it deforms, it strains, it loads. TFM therefore lets us watch cells not only as decision-makers responding to cues, but as agents that *build and remodel* their environments through physical work.

Under the microscope, a migrating cell can look like a delicate, almost fragile thing. TFM reveals it as an active mechanical player: a creature that grabs, tests, and pulls, leaving behind a transient pattern of deformations like footprints in fresh snow that melt as soon as the foot lifts. In a living tissue those footprints are not merely erased; they can be remembered by the matrix itself. Fibres can be reoriented, cross-links can be formed, and the mechanical landscape can be permanently altered. The cell's invisible grip, once mapped, becomes part of a larger story of how tissues are sculpted by the forces of their own inhabitants.

## 9.2 The Atomic Finger

A traction map tells you what a cell *does* to its surroundings: where it pulls, how hard, and how that pattern changes as it crawls. But there is another question, equally fundamental and sometimes more immediate: what does the cell *feel*? If a cell is a climber, it does not only grip; it also tests the rock. It probes for firmness, for give, for the difference between a stable hold and crumbling sand. In tissues, those differences are not cosmetic. The stiffness of the environment can decide whether a stem cell settles into bone or fat, whether an immune cell slips through a vessel wall, whether a cancer cell finds the courage to invade.

Human fingers are exquisitely tuned to stiffness. A clinician can palpate a swollen lymph node and learn something in seconds that would take longer to explain in words: this is rubbery, that is hard, the border is smooth, the texture is irregular. Mechanobiology wanted an instrument that could do the same to cells and their microenvironments, but on a scale a thousand times smaller, with a readout that becomes a number rather than a hunch. Atomic Force Microscopy (AFM) became that instrument: a blind finger that can press on a living cell, feel its resistance, and translate that resistance into a stiffness value.

Despite its name, AFM is not mainly about atoms in this context. The "atomic" label reflects its origins in surface science, where the technique was developed to image surfaces at extremely high resolution. The leap to biology came when researchers realized that the same cantilever that traces the bumps of a crystal can also indent a cell membrane without tearing it, and that the bending of that cantilever can be measured with astonishing sensitivity. Instead of reading out a landscape of height, AFM can read out a landscape of mechanical compliance: how much a sample yields when pressed.

The heart of the instrument is a cantilever, a microscopic diving board anchored at one end. At the free end sits a tip: sometimes a sharp pyramid, sometimes a small bead glued on like a rounded knuckle. A laser reflects off the back of the cantilever into a photodetector. When the tip touches a surface and the cantilever bends, the reflected laser beam shifts. That shift is converted into force using a simple principle: the cantilever is itself a calibrated spring. If you know its spring constant—how many newtons of force are required to bend it a given distance—then its deflection tells you the force applied at the tip.

Force is only half the story. The other half is indentation: how far the tip pushes into the sample. Combine the two and you obtain a force-indentation curve, a mechanical fingerprint. Press lightly and some materials yield easily; press harder and they may stiffen. Release and they may spring back or lag behind. Cells, being living and structured, often do all of the above. Their outer membrane is thin and slippery, but it is supported by a cortex of actin filaments under tension. Beneath that is a crowded interior with organelles, a nucleus that can feel like a stone in a pillow, and a cytoskeleton that can rearrange while you are pressing on it.

The number mechanobiologists often want from AFM is *Young's modulus*, a standard measure of stiffness. Roughly speaking, it tells you how much stress (force per area) is needed to produce a given strain (fractional deformation). For an ideal, uniform, purely elastic material, Young's modulus is a single property like density. For cells and tissues, it is more like a summary statistic. Living matter is not uniform and it is not purely elastic: it flows, remodels, and actively generates tension. Even so, Young's modulus is useful because it anchors a discussion. It lets one say, with some precision, that one sample is ten times stiffer than another, or that a drug softened a cell by half.

How does a tiny cantilever convert a force curve into Young's modulus? The conversion relies on contact mechanics: the mathematics of how objects deform when they touch. The simplest model treats the sample as an elastic half-space and the indenter as a known shape. The Hertz model, developed in the nineteenth century for spheres pressing on elastic materials, often appears in AFM analysis. If the AFM tip is approximated as a sphere (or if a bead is used deliberately), then force increases with indentation in a predictable way. Fit the measured curve to the model and you can solve for Young's modulus. If the tip is sharp, other models apply, but sharp tips bring their own hazards: they concentrate force on a very small area and can puncture membranes or probe only the superficial cortex rather than the bulk.

The craft of AFM in biology lies in understanding what you are *really* measuring. Press on a cell and you are not measuring "cell stiffness" in the way you measure the stiffness of a rubber block. You are measuring a combination of cortical tension, cytoskeletal architecture, internal pressure, and viscous response, all filtered by the geometry of the indenter and the timescale of indentation. Press quickly and the cell may seem stiff because fluid cannot redistribute fast enough. Press slowly and the same cell may appear softer because the cytoskeleton relaxes and water shifts. Even the temperature and chemical environment matter, because cells actively regulate their tension. A cell is not merely a material; it is a machine that adjusts its own settings while you poke it.

That active nature is exactly why AFM became so influential. It can measure mechanics in conditions close to life: cells in culture medium, at body temperature, responding to cues. It can map mechanical variation across a single cell by indenting many points in a grid: the periphery may be softer than the region above the nucleus, and the stiffest zones may correlate with stress fibres or adhesion sites. It can also compare cell types. One of the most intuitive contrasts is between cells from soft tissues and those from hard ones. Neurons and glial cells associated with the brain tend to be remarkably compliant; osteoblast precursors associated with bone tend to be much stiffer. These differences are not mere consequences of where the cells live; they are part of what those cells *are*, tied to their cytoskeletal composition and the mechanical demands of their roles.

A concrete example makes the stakes clearer. Consider red blood cells. Their job is to squeeze through capillaries narrower than the cell's own diameter. That requires deformability. In diseases such as malaria, infected red blood cells become stiffer, and that stiffness contributes to capillary blockage and severe complications. AFM has been used to measure changes in red blood cell stiffness at the level of single cells, revealing mechanical signatures that track infection stage. The point is not that AFM is a routine diagnostic tool in the clinic, but that it exposes a mechanistic link: a pathogen can harm by altering the mechanical properties of its host cell, turning a flexible traveller into a rigid clog.

Cancer provides another vivid context. Tumours are often palpably stiff, a fact known long before mechanobiology had its name. A hard breast lump is one of the oldest warning signs. Part of that stiffness comes from the surrounding matrix: collagen fibres laid down and cross-linked, tissue architecture reorganized, fluid pressure elevated. But cancer cells themselves can also change mechanically, sometimes becoming softer to squeeze through tight spaces, sometimes stiffening certain regions to generate invasive forces. AFM has helped disentangle these layers by separately measuring the stiffness of cancer cells and of the matrix they inhabit. That separation matters, because therapies that target the tumour microenvironment might aim to soften the matrix, while therapies that target cytoskeletal regulation might alter cell mechanics directly. In both cases the central idea is the same: mechanics is not an afterthought; it is a controllable parameter that shapes disease.

AFM's historical arc adds a touch of drama. The technique emerged in the 1980s, in the same intellectual family as the scanning tunnelling microscope that could "see" atoms on conductive surfaces. AFM's genius was to replace electrical tunnelling with physical force, allowing imaging of non-conductive samples. That opened the door to polymers, ceramics, and eventually biological specimens. In the early days, the thought of dragging a probe across something as delicate as a cell membrane sounded like vandalism. Yet with gentle forces and careful control, AFM could do something optical microscopy could not: measure not only what cells look like, but how they resist being deformed.

The sensation of *palpation* is a helpful metaphor, but it hides the precision. A typical AFM indentation experiment might apply forces in the range of piconewtons to tens of nanonewtons. At the low end, that is comparable to the force a molecular motor can generate. At the high end, it is enough to noticeably deform a cell but still far below what would tear most cells if applied smoothly and over a broad tip. The indentation depth might be tens to hundreds of nanometres for cortical probing, or up to a micron or more for deeper mechanical response. Those numbers matter because cells are layered. A shallow indentation largely interrogates the actin cortex and membrane tension; a deeper indentation begins to sense the cytoplasm and, if you press over it, the nucleus.

AFM also encourages a more careful vocabulary. Stiffness is often used casually to mean "hardness," but hardness usually refers to resistance to permanent deformation or scratching, while stiffness refers to elastic resistance. Tissues are not hard in the way bone or enamel is hard; they are stiff or soft, and they are often viscoelastic. Viscoelasticity means the response depends on time. Press into cartilage and it rebounds slowly; stretch a tendon and it relaxes over minutes. Cells show similar behaviour on shorter timescales. AFM can quantify this by measuring how force changes under a constant indentation (stress relaxation) or how indentation changes under a constant force (creep). The curves become signatures of internal rearrangement: cytoskeletal remodelling, fluid redistribution, bond unbinding and rebinding. This is one reason AFM remains valuable even when a single Young's modulus seems too crude. The full time-dependent response can reveal whether a drug altered actin crosslinking, myosin activity, or membrane tension.

There is, however, a trap that catches the unwary: the numbers depend on assumptions. The Hertz model assumes a homogeneous, isotropic, purely elastic sample with a flat surface and small strains. A living cell violates nearly every clause of that sentence. If the indentation is too deep, the underlying glass dish may contribute to the apparent stiffness, making the cell seem stiffer than it is. If the tip is sharp, you may be measuring the cortex rather than the cytoplasm. If the cell is actively contracting, the force curve can include contributions from active tension, not merely passive resistance. Skilled practitioners therefore treat AFM stiffness not as an intrinsic constant but as a measurement made under specified conditions: tip geometry, indentation depth, loading rate, location on the cell, temperature, and whether the cell is adhered and spread or rounded and suspended.

These constraints, rather than diminishing AFM, make it scientifically honest. In mechanobiology, every measurement is a negotiation between what one wishes to know and what one can control.

AFM's strength is that it makes the negotiation explicit. It tells you exactly what force was applied, exactly how far the sample yielded, and at what speed, and it does so with spatial specificity. That specificity lets it answer questions that traction maps alone cannot. A cell might exert strong tractions because it has a robust actomyosin system, but it might still be mechanically soft if its cortex is weak. Conversely, a cell might be stiff yet exert modest tractions if its adhesions are immature. Force generation and mechanical resistance are related but not identical properties. AFM measures the latter directly.

AFM also sits within a broader family of force-measuring tools that span scales. Optical tweezers can trap and tug on beads with piconewton forces; magnetic tweezers can apply constant forces and torques; AFM can apply higher forces and measure with high bandwidth. At the molecular end, biophysicists have used these tools to unfold proteins, stretch DNA, and watch motors step along filaments. Those experiments teach something that is psychologically useful for cell mechanics: piconewtons are not negligible. A few piconewtons per molecule, multiplied across thousands of molecules in an adhesion, becomes a macroscopic action that can reorganize tissue. The bridge between molecule and cell is built out of many small forces acting in parallel.

That bridge has been strengthened by molecular tension sensors that report piconewton loads inside living cells. A particularly influential example used a genetically encoded sensor inserted into vinculin, a key protein in focal adhesions that links integrins to the actin cytoskeleton. By designing a spring-like module whose extension changes fluorescence through a phenomenon called Förster Resonance Energy Transfer (FRET), researchers could infer the tension borne by vinculin molecules in different adhesion states. These sensors add a missing layer: they indicate not just what force a cell exerts on a substrate or how stiff it feels under a probe, but what forces specific proteins experience while doing their jobs. The conceptual twist is that localization does not guarantee load-bearing: a protein can be present in an adhesion without carrying much tension, and tension can change without dramatic changes in protein abundance. Mechanics is not only about where molecules are; it is about what they are *doing*.

DNA-based tension probes take a different, almost playful approach. They use DNA hairpins that unfold at particular force thresholds, producing a switch-like fluorescence signal when an attached receptor pulls hard enough. Rather than yielding a continuous force value, they report whether local forces exceed a chosen threshold—a kind of mechanical "yes/no" map. This is reminiscent of old photographic film that either exposes or does not, depending on light intensity, except the currency is force at the level of single receptors.

Why mention these molecular tools while talking about AFM? Because they clarify what AFM numbers mean. When AFM reports that a cell's apparent Young's modulus changed, the real question is which physical elements changed. Did the cortex tense? Did actin crosslinking increase? Did osmotic pressure shift? Did the cell simply spread more, placing the probe over a different internal landscape? Molecular tension sensors and traction microscopy can provide complementary answers, but AFM's distinctive contribution is palpation: it probes the mechanical response directly and locally, in a way that resembles how cells themselves probe their surroundings.

A satisfying way to picture AFM is as an experiment in reciprocity. The cell presses on the world and the world presses on the cell. With traction microscopy one listens to the cell's outward push and pull. With AFM one pushes back and asks, gently but insistently, "how do you respond?" The response is not merely passive compliance. Cells stiffen when they activate contractility, soften when their cytoskeleton is disrupted, and reorganize under sustained load. In the living body those responses are happening constantly: endothelial cells lining blood vessels feel shear stress from flowing blood; cartilage cells feel compressive loads; immune cells squeeze and pull through dense matrix. AFM cannot recreate the whole theatre, but it can isolate the mechanical dialogue into a controlled, measurable exchange.

There is a final, human dimension to this "atomic finger." Many of the processes that most affect our lives—development, wound repair, scarring, tumour progression—depend on mechanical properties that clinicians have long sensed with their hands but struggled to quantify. AFM is one of the instruments that converts that tacit knowledge into a language of numbers, curves, and maps. In doing so, it gives researchers a way to connect the felt qualities of living tissue—soft, stiff, yielding, tense—to specific cellular structures and molecular mechanisms, and to treat mechanics as something that can be measured, compared, and ultimately engineered.

## 9.3 Commanding the Force

Measurement is a kind of restraint. A cell is allowed to behave, and the scientist watches closely, trying not to disturb the scene. That attitude has produced exquisite maps of traction and careful measurements of stiffness, yet it leaves a stubborn doubt: are the forces *causing* the biology, or are they merely a side effect of something else? A cell that spreads on a stiff surface also changes gene expression, metabolism, and signalling. If you perturb a gene and the traction forces drop, perhaps force matters—or perhaps the cell is simply unwell. Causation is hard to earn in a living system.

Mechanobiology has learned to earn it by learning how to *command* force: to dial contractility up and down on demand, to apply a mechanical input at a chosen receptor, or to build an environment that quietly forces the cell to make a different mechanical bargain. These manipulations are not tricks; they are the equivalent of the experimental controls that chemistry and genetics take for granted. If a cell's behaviour changes when you impose a force change with precise timing and location, the force is no longer a correlation. It has become an intervention.

One route to command comes from light. The appeal is obvious: light can be delivered with exquisite spatial precision, can be switched on and off in milliseconds, and can be aimed at a single corner of a single cell without touching it. Biology already uses light as information—vision, circadian rhythms, photosynthesis—so it is not surprising that proteins exist that change shape when they absorb photons. Optogenetics repurposes those proteins. It began as a way to control electrical activity in neurons with light, but the underlying idea is broader: if a protein can be activated by illumination, any pathway downstream of that protein becomes light-controllable.

In mechanobiology, the pathway of choice is often the one that controls actomyosin contractility: the cell's internal muscle. Actin filaments form tracks and scaffolds; myosin motors pull on them, generating tension. A central switch for this machinery is the small GTPase RhoA, which regulates downstream effectors such as ROCK (Rho-associated kinase), increasing myosin activity and stabilizing actin structures. When RhoA signalling rises, cells typically become more contractile: they pull harder on adhesions, stiffen their cortex, and often rearrange their shape. If you can activate RhoA with light, you can produce a controlled surge of force.

The practical trick is to use a light-sensitive interaction that recruits an activator of RhoA to the cell membrane, where it can act efficiently. One family of designs relies on engineered pairs of proteins that bind only when illuminated, like a molecular handshake triggered by blue light. Another approach has a pleasing simplicity: a single-component system in which a light-sensitive domain changes its affinity for the membrane, bringing along a RhoA activator as cargo. When the light turns on, the activator rushes to the membrane; when the light turns off, it disperses. Because the membrane is a stage where many signalling events are organized, this recruitment can rapidly reshape the cell's mechanical state.

The emotional punch of optogenetics is the moment the cell responds. A patch of light falls on a region of the cell edge, and within seconds the cytoskeleton tightens there, adhesions strengthen, and traction forces rise. If traction force microscopy is running in parallel, the beads in the gel report the change as a new bloom of displacement. You can draw an invisible circle around the illuminated region and watch the traction vectors rotate, intensify, or concentrate. The cell becomes, briefly, a mechanical instrument you can play: a local chord of contractility here, a wave of tension there.

This kind of control has clarified a long-standing puzzle about how cells polarize during migration. A crawling cell needs a front and a back. Biochemistry provides many polarity cues, but mechanics can reinforce them: stronger contraction in one region can stabilize adhesions and pull the cell body forward, while weaker contraction elsewhere allows protrusion. By activating RhoA locally, one can test whether a mechanical asymmetry is sufficient to bias motion. The answer, in many cell types, is that it often is. A cell exposed to localized contractile activation can turn, retract a protrusion, or redirect its leading edge. The light does not supply a chemical gradient in the medium; it supplies a gradient of *tension* inside the cell.

The story becomes even more compelling when force is coupled to mechanotransduction readouts, the molecular events through which cells convert mechanical inputs into biochemical signals. One well-studied mechanotransduction axis involves the transcriptional regulators YAP and TAZ, which shuttle between cytoplasm and nucleus depending on mechanical context. On stiff substrates, or when the cytoskeleton is highly tensed, YAP tends to accumulate in the nucleus and drive gene expression programs associated with growth and certain differentiation states. When contractility is reduced, YAP often exits the nucleus. Optogenetic control of RhoA can therefore be used not merely to make a cell pull harder, but to test whether an induced pulse of contractility is enough to push YAP into the nucleus. When it is, the causal chain becomes palpable: light $\rightarrow$ RhoA activation $\rightarrow$ increased tension $\rightarrow$ altered nuclear signalling. The route from mechanics to gene regulation is no longer inferred; it is driven.

Light-based command has a limitation that is also a virtue: it manipulates the cell's own machinery. That means it respects, and depends on, the cell's internal organisation. If a pathway is missing, inhibited, or rewired by disease, the light switch may do little. Sometimes mechanobiologists want a more blunt instrument: a way to apply force *from the outside*, regardless of the cell's internal signalling, and preferably targeted to a particular receptor.

Magnetic actuation provides that instrument. The basic idea can be explained with a toy: sprinkle iron filings on paper, place a magnet beneath, and the filings move, rotate, align. Replace the filings with micron-sized magnetic beads attached to a cell surface receptor, and replace the hand-held magnet with a calibrated electromagnetic setup, and you have a way to apply controlled forces and torques directly to the cell's surface.

The beads can be coated with ligands that bind specific receptors, such as integrins, which are the cell's main mechanical link to the extracellular matrix. When an integrin-bound bead is pulled or twisted by a magnetic field, the receptor experiences a force, and so does the intracellular machinery attached to it. This approach has a history in the form of magnetic twisting cytometry, where oscillating magnetic fields twist beads and the cell's resistance is measured. But its deeper value is not only measurement; it is perturbation. A bead can be tugged steadily to mimic a persistent load, or pulsed to mimic a transient mechanical event. It can be applied at one point or many points. It can even be used as a mechanical "question": if you pull here, does the cell respond there?

A concrete example brings the point home. Imagine two beads on the surface of the same cell. One is bound to integrins; the other is bound to a receptor that is not mechanically coupled to the cytoskeleton. Apply the same magnetic twist to both. The integrin-bound bead often elicits a cellular response: reinforcement of the actin network near the bead, recruitment of focal adhesion proteins, activation of signalling pathways. The control bead may rotate more freely and provoke little response. The difference is not in the magnet; it is in the mechanoreceptor. By choosing what the bead binds, magnetic actuation can test whether a particular receptor is capable of transmitting force into the cell in a biologically meaningful way.

Magnetic actuation also makes a philosophical point about mechanobiology: force is not a generic stimulus like heat. It is *delivered through structures.* A mechanical load applied through integrins engages focal adhesions; a load applied through cadherins engages cell–cell junctions; a load applied to the membrane more generally engages membrane tension and associated channels. The same magnitude of force can have different meanings depending on the route by which it enters. Magnetic beads let one choose the route.

Optogenetics and magnetic actuation share a crucial property: timing. The cell is not a static material; it is a dynamic system with feedback. Pulling for one second and pulling for one hour are different interventions. A brief pulse can reveal rapid mechanosensitive signalling, while sustained loading can reveal adaptive remodeling:

reinforcement, softening, changes in gene expression. With precise timing, one can test whether a mechanotransduction pathway is fast (seconds), intermediate (minutes), or slow (hours). That temporal fingerprint helps identify the mechanisms involved: ion channels and phosphorylation are fast; transcriptional responses are slower; matrix remodeling and differentiation are slower still.

The third way to command force is more subtle, and it requires no lasers and no magnets. It is done by building environments that impose mechanical constraints by their very nature. Cells are exquisitely sensitive to their substrate: its stiffness, its thickness, its ability to relax, its topology, even the spacing of adhesion sites. By engineering the substrate, one can compel cells to adopt certain force regimes and then ask what follows.

Hydrogels are the workhorses of this approach. A hydrogel is a network of polymers swollen with water, like a microscopic sponge. By changing the polymer concentration or cross-linking density, one can tune its stiffness across orders of magnitude, from brain-soft to cartilage-firm. Because the gel is mostly water, it is friendly to cells; because it is a network, it can be designed with chemical handles to attach adhesion proteins. The important point is that stiffness can be changed without changing the chemistry of the ligands the cell binds. This separation is precious, because it avoids a common confound: in biology, changing one variable often changes many.

Engineered hydrogels can do more than set a single stiffness. They can present stiffness gradients, allowing a cell to crawl from soft to stiff regions, a behaviour known as durotaxis. They can incorporate viscoelasticity, letting the gel relax over time like real tissues do. They can be made degradable, so cells can carve into them, creating a moving boundary between cellular force and material resistance. Each design choice is a way of asking a question in the language of mechanics.

One of the most striking behavioural contrasts produced by stiffness control is the difference in cell shape. On very soft matrices, many cells remain rounded or only partially spread, with small, transient adhesions and weak traction. On stiff matrices, the same cells often spread widely, build prominent stress fibres, and generate large tractions. It is tempting to describe this as "the stiff matrix activates the cell," but the more accurate description is a negotiation: the cell's contractile machinery tests the substrate, and when the substrate resists, the cell can build tension, stabilize adhesions, and reinforce its cytoskeleton. When the substrate yields, tension cannot build as effectively, adhesions slip, and the cell settles into a different state.

This negotiation has deep consequences. In stem cell biology, stiffness has been used as a cue that biases differentiation. In fibrosis, pathological stiffening of tissue can lock cells into a contractile, matrix-producing phenotype, creating a feed-forward loop: stiffer matrix promotes more contraction and more matrix deposition, which stiffens the tissue further. Engineered hydrogels let one reproduce pieces of that loop in controlled settings and test where it can be broken. A drug that interrupts contractility might soften the mechanical feedback even if it does not alter the matrix chemistry. Conversely, therapies aimed at reducing cross-linking in the matrix might lower stiffness and thereby reduce cellular activation.

Commanding force through environment engineering also extends beyond flat surfaces. Cells in the body live in three-dimensional matrices, not on two-dimensional planes. In 3D, the rules change. Adhesions are distributed around the cell, not concentrated on a basal surface. Protrusions push into a fibrous mesh. The matrix can be remodeled, aligned, and densified. Engineering 3D hydrogels and collagen networks makes it possible to ask how traction and stiffness sensing operate when forces are distributed in all directions and when the cell can physically rearrange its environment.

A historical anecdote captures how radical this perspective once seemed. For much of twentieth-century cell biology, the standard culture surface was glass or rigid plastic, chosen for convenience and clarity rather than physiological relevance. Cells were asked to behave on something closer to a countertop than to living tissue. When researchers began to culture cells on compliant gels and observed dramatic changes in shape, motility, and gene expression, it was not merely a new method; it was an argument that many "normal" behaviours in culture were actually artefacts of unnatural stiffness. The shift resembles an earlier revolution in microbiology, when it became clear that bacteria behave differently in biofilms than in free-floating culture. In both cases, the environment was not a backdrop. It was a participant.

What makes these three command approaches powerful is how readily they can be combined with the measurement tools already in hand. Turn on optogenetic RhoA activation and watch traction vectors intensify in real time. Twist a magnetic bead and measure the local stiffening of the cortex with AFM. Place a cell on a viscoelastic hydrogel and ask whether its traction pattern changes not only in magnitude but in persistence. Layered together, they allow experiments that feel almost like conversations with the cell: a mechanical question posed, a mechanical answer recorded, and then a second question that tests whether the first answer mattered.

There is also a humility built into the best of these manipulations. Commanding force does not mean overriding the cell like a puppet. It means perturbing a system that is already active and adaptive, and then learning from how it compensates. A cell may respond to imposed tension by reinforcing adhesions, or by yielding and flowing, or by changing gene expression to remodel its cytoskeleton. Those responses are not annoyances; they are the substance of mechanobiology. They are how living matter maintains integrity under load, how tissues heal, and how disease sometimes hijacks the same machinery for destructive ends.

A final image is worth keeping in mind: a single cell embedded in a tissue, surrounded by fibres, neighbours, and fluid. Forces arrive from many directions—blood flow, muscle contraction, the pull of development, the pressure of swelling. The cell's own forces push back, not as a simple reaction but as a regulated, purposive act. To command force in the laboratory is to recreate, in miniature and with intent, the mechanical arguments that cells endure and negotiate every day, and to watch what they become when those arguments change.

Chapter 10

# Healing with Physics: The Future of Mechanomedicine

*For centuries, medicine has viewed the body primarily as a chemical vessel, treating ailments with pills and compounds. This chapter reveals a paradigm shift toward 'mechanomedicine,' where calibrated forces and engineered environments recruit the body's own physical sensors to heal wounds, regenerate organs, and perhaps rewrite the definition of biological life.*

## 10.1 Medicine Beyond Molecules

Modern medicine is often narrated as a triumph of chemistry: find the right molecule, match it to the right target, and let a biochemical lock click shut around its key. The story is true, and its successes are extraordinary. Yet it is incomplete in a way every person with a pulse already knows. Your heart does not beat because a ligand binds a receptor. Your lungs do not inflate because a drug diffuses into an enzyme pocket. Life runs on forces: pressure, stretch, shear, compression. Cells live in a mechanical world and, crucially, they *listen* to it.

That listening is not mystical. It is molecular. The same cell that responds to a chemical message can respond to a physical one, and sometimes it does so using the same internal wiring. The clinic has always taken advantage of this in a rough-and-ready way. Surgeons reduce fractures and immobilize them. Physiotherapists prescribe exercise. Cardiologists measure blood pressure. But these practices have often treated mechanics as background conditions: important, certainly, yet too blunt to be considered a precise signal. Mechanotherapy and mechanomedicine argue for something sharper: force as an input that can be *dosed*, *timed*, and *shaped* to trigger specific cellular pathways, much as drugs do.

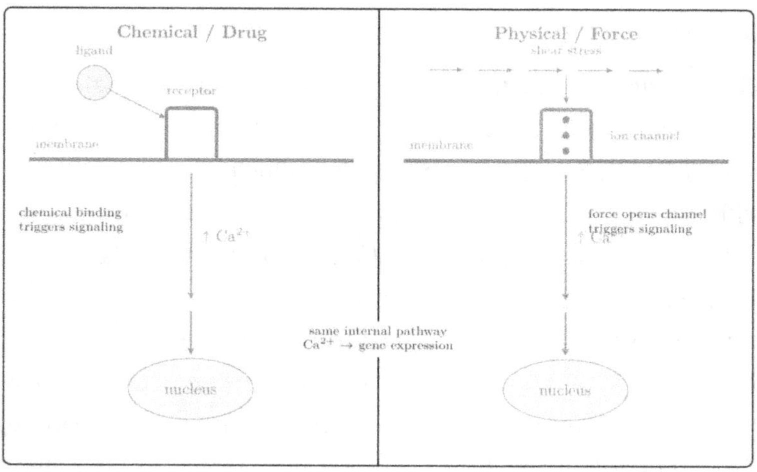


One reason this equivalence matters is that cells are not passive recipients of their environment. A chemical message arrives, is interpreted, and fades. A mechanical message can be *made* by the cell itself and then amplified in a loop. A fibroblast in a healing wound pulls on its surroundings; if the surrounding matrix is stiff, that pull generates higher internal tension, which pushes the fibroblast toward a more contractile, scar-forming state. The cell then secretes and crosslinks more matrix, making the environment even stiffer. In such a loop, mechanics is not a side effect. It is part of the disease process.

To see why mechanotherapy can be precise, it helps to picture the cell as a small tent. The membrane is the canvas; the cytoskeleton is the poles and ropes; adhesions are the pegs. A drug might act by changing the chemistry of a peg or a rope. A force acts by tugging on the whole structure. But tugging does not have to be crude. The direction of pull, its magnitude, the frequency of the tugging, and the moment you apply it can each select different molecular outcomes. A high-frequency vibration is not the same as a slow stretch. A brief pulse is not the same as sustained tension. Even shear stress (the frictional drag of a fluid sliding past a surface) can be a finely tuned language, especially for cells lining blood vessels.

Consider the endothelium, the thin layer of cells that coats every artery and vein. These cells constantly experience shear from blood flow. When flow is smooth and steady, endothelial cells tend to align with the direction of flow and adopt an anti-inflammatory, protective state. Where flow is disturbed—at branch points or bends—the cells receive a more chaotic mechanical message and can become more inflammatory, encouraging plaque formation. Here the same body demonstrates two physiological states not because of different DNA, but because of different *physics*. The cell reads the message through mechanosensitive structures: ion channels that open under tension, adhesion complexes that change shape when pulled, and a cytoskeleton that routes forces toward the nucleus, where gene expression can be reshaped.

A natural skepticism arises: if mechanics is so fundamental, why did medicine become so molecule-centric? Part of the answer is practical. Molecules are easy to bottle, ship, and prescribe. Forces are harder. They seem to belong to the domains of surgery, rehabilitation, and devices, not to the clean world of pills. Another reason is measurement. Chemistry comes with concentration units, receptor affinities, and pharmacokinetics. Mechanics can feel like a messy collection of qualitative descriptors: "gentle," "vigorous," "weight-bearing," "rest." Mechanomedicine tries to turn those descriptors into controlled parameters: strain (how much something deforms), stress (force per area), stiffness (how hard it is to deform), and time.

Once you accept that force can be dosed, the clinic looks different. Physical therapy is no longer a set of folk practices refined by experience; it becomes a kind of pharmacology of loading. One of the earliest and most enduring observations in orthopedics is that bone adapts to mechanical demands. This idea is often associated with Julius Wolff in the nineteenth century, and although the details have been revised, the central fact stands: bone is a living material that remodels in response to load. Too little load and it weakens; the extreme example is the rapid bone loss seen in prolonged bed rest or microgravity. Appropriate load and it strengthens; this is why weight-bearing exercise is protective against osteoporosis. The remarkable step is to translate this general truth into specific prescriptions after injury: not simply *move* or *rest*, but load in the right way at the right time.

A historical anecdote makes the point vivid. In the 1950s, the Soviet surgeon Gavriil Ilizarov developed a method for repairing complex fractures and lengthening limbs using an external ring fixator. The apparatus looks like scaffolding: rings around the limb connected by rods and wires that hold bone fragments in place. Ilizarov's key insight was not just mechanical stabilization. He used controlled, gradual distraction—slowly separating bone segments by small daily increments—to stimulate new bone formation in the gap. The body responds to this mechanical environment by generating tissue, not merely by sealing a break. Today the method is used worldwide, refined and modernized, and its logic is mechanobiological: the patient's cells interpret a sustained mechanical regimen and build accordingly.

This is mechanotherapy in a pure form: a physical input designed to produce a biological output, with parameters that can be adjusted. Too fast a distraction, too much tension, or poor alignment and the tissue response fails. Done well, it produces bone where none existed. The fixator is a device, but the deeper lesson is conceptual: the therapeutic agent is *mechanical information.*

Not every mechanotherapy requires dramatic hardware. Some of the most exciting versions are subtle: low-intensity pulsed ultrasound for certain fractures, extracorporeal shockwave therapy for stubborn tendon problems, or carefully structured loading programs for tendinopathy and ligament rehabilitation. These approaches can sound like "just physics," as if the body were being shaken into compliance. Yet their rationale is increasingly specific. Tendons, for example, are built to transmit force and are populated by cells that live under tension. After injury, tendon tissue can become disorganized, painful, and mechanically compromised. Controlled loading can encourage collagen alignment and appropriate remodeling; complete rest often backfires by depriving cells of the very signals that maintain tendon structure. The therapeutic trick is to provide enough strain to activate beneficial remodeling pathways without tipping into overload and re-injury.

The idea of a mechanical "dose" becomes clearer when you think about what a cell can measure. A cell cannot hold a dynamometer. But it can sense membrane tension, the stretching of protein domains, the opening probability of an ion channel, and the rate at which adhesions assemble and disassemble. These are molecular events with thresholds and kinetics. A mechanosensitive ion channel may open when membrane tension exceeds a certain range, letting calcium ions rush in. Calcium is not only a mineral for bones; inside cells it is an electrical-chemical messenger, a fast way to translate an external event into a cascade of enzymes and gene regulation. In that sense, a shear stress that opens a channel and a ligand that activates a receptor can converge on the same intracellular currency.

Calcium is also a reminder that "mechanical" and "chemical" are not competing explanations. They are interlocked. A force can release a growth factor from a matrix, alter how receptors cluster, or change how quickly molecules diffuse. Conversely, a drug can change the stiffness of a tissue by altering collagen crosslinking, inflammation, or cell contractility. Mechanomedicine is not an attempt to replace pharmacology, but to broaden it: to treat the mechanical state of tissue as both a cause and a target.

Nowhere is this broader view more relevant than in fibrosis and scarring. In many organs—lung, liver, heart, kidney—chronic injury can lead to excessive matrix deposition that stiffens tissue and impairs function. Once stiffness increases, cells that specialize in producing matrix (often myofibroblasts) become more activated, pull harder, and produce still more matrix, driving a self-reinforcing cycle. Chemical signals like TGF-$\beta$ are deeply involved, but mechanics is part of the switch that makes those signals decisive. The result is a disease that is partly biochemical and partly mechanical: a "hardware" problem and a "software" problem at the same time.

A mechanotherapeutic approach to such conditions can take several forms. One is to target the mechanical environment indirectly by preventing matrix stiffening (for example, by interfering with enzymes that crosslink collagen). Another is to alter how cells sense stiffness, by targeting mechanotransduction pathways that translate force into gene expression. But there is also a more literal possibility: to deliver mechanical cues that reprogram tissue behavior. That might sound far-fetched until you remember that organs are already mechanotherapeutic devices. Every breath stretches lung tissue rhythmically.

Every heartbeat strains the cardiac wall. Every step loads cartilage, bone, tendon, and muscle. Disease can distort these rhythms, and therapy can sometimes restore them.

Wound healing offers a particularly intimate example, because it is visible and personal. A cut on the skin closes partly because cells migrate and contract, pulling the wound edges together. The cells doing the pulling respond to tension: too little, and closure is sluggish; too much, and scarring can be exaggerated. Clinicians have long used mechanical ideas without naming them as such. Sutures are placed not only to approximate edges but to distribute tension. Compression garments are used after burns to reduce hypertrophic scarring. Silicone sheets and taping techniques are used to modulate the mechanical state of healing skin. These are not merely passive supports. They are attempts to tune the forces experienced by cells during a critical window, when those forces influence which genes are expressed and which tissue architecture emerges.

The phrase "critical window" matters. A drug taken too early or too late can miss its effect because the relevant pathways are not active. Mechanical cues can be just as time-sensitive. Early after injury, inflammation dominates; later, proliferation and matrix deposition take over; later still, remodeling reshapes the tissue. Loading a tendon aggressively on day one is not the same as loading it gently after the initial inflammatory surge. Immobilizing a joint briefly can protect a fragile repair; immobilizing it for too long can lead to stiffness, muscle atrophy, and poor tissue quality. Mechanotherapy, at its best, treats timing as part of the prescription.

Precision also means personalization. Two patients can receive the same exercise program and experience different tissue strains because their anatomy, muscle strength, and movement patterns differ. A mechanomedicine mindset pushes toward measurement and feedback: sensors that quantify gait loading, braces that deliver controlled stress, rehabilitation protocols that adjust in response to objective signs of tissue adaptation. Even simple technologies—accelerometers, pressure insoles, wearable heart-rate monitors—hint at a future where physical "dosing" becomes as trackable as medication adherence.

An underappreciated feature of force-based therapies is that they can, in principle, act locally without requiring systemic exposure. A drug swallowed for a bone problem circulates through every organ; side effects are often the price of access. A mechanical input can sometimes be targeted: ultrasound focused on a fracture site, shockwaves aimed at a calcified tendon insertion, or a stretching regimen applied to a specific muscle-tendon unit. This does not make mechanotherapy risk-free. Misapplied forces can injure. But it changes the calculus of precision: the therapeutic signal can be confined to the tissue where it is needed, and its parameters can be adjusted in real time.

The deeper promise of mechanomedicine is not simply "more devices" or "better rehab," though those will come. The promise is a conceptual unification: chemistry and physics as two dialects that cells translate into the same internal language. In that language, a receptor's conformational change and an ion channel's opening are both ways of turning an external event into intracellular currents of calcium, phosphorylation cascades, cytoskeletal remodeling, and ultimately gene expression. When that translation goes wrong, disease can follow. When it is steered carefully, healing can be accelerated.

A person recovering from a fracture or a tendon tear often asks a deceptively simple question: how much should I move, and when? For a long time, the honest answer was "it depends," backed by experience, tradition, and caution. Mechanotherapy does not eliminate that complexity, but it makes the dependency legible. The body is not merely repairing a structure; it is running a mechanical experiment on itself, sampling forces and updating tissue architecture in response. The clinician's role, increasingly, is to shape those forces so the experiment converges on function rather than on pain, stiffness, or scar. When physics is treated as a controllable signal rather than an unavoidable stress, rehabilitation stops being the afterthought to medicine and becomes one of its most exacting forms.

## 10.2 Architects of Living Tissue

A surgeon can stitch a wound closed, but no human hand can stitch collagen fibres into the precise weave of a tendon, or lay down the microscopic honeycomb of bone. That work is delegated to cells. The dream of regenerative medicine is to give those cells the right instructions so they rebuild what disease or injury has taken away. For a long time, the instruction manual was assumed to be chemical: growth factors, hormones, cytokines, small molecules. Those certainly matter. Yet stem cells—the body's great reserve of potential—also make decisions by touch. Give them a surface that feels like brain and they behave one way; give them one that feels like bone and they behave another. It is as if the cell presses a thumb into its surroundings, judges the give, and chooses a future.

This is not metaphor. Cells literally push and pull. A stem cell spreads on a surface by grabbing it with protein "hands" called integrins, which span the membrane and latch onto extracellular proteins (or, in the lab, onto adhesive coatings chosen by the experimenter). Inside the cell, integrins connect to a scaffolding of actin filaments and molecular motors. Those motors, chiefly myosin, slide filaments past each other and generate tension, much the way muscle does on a smaller scale. A cell on a firm surface can pull and feel resistance; on a soft one, its tug simply deforms the material and the cell feels less internal strain. That difference in internal strain is converted into molecular signals that reach the nucleus and influence which genes are turned on.

The materials that surround cells in living tissue—the extracellular matrix, or ECM—are not inert packaging. They are a hybrid of chemistry and mechanics: collagen and elastin fibres, proteoglycan gels, bound growth factors, water, and a constantly changing architecture. Tissue engineers borrow the idea and build simplified versions: hydrogels that can be tuned from jelly-soft to cartilage-like, fibrous meshes that mimic tendon, porous scaffolds that allow blood vessels to infiltrate. These are not mere supports, like rebar in concrete. They are intended as active cues, a kind of physical grammar that cells can read.

A turning point came when researchers began to show, in a controlled way, that substrate stiffness alone could bias stem cell fate. In 2006, Adam Engler and colleagues reported that mesenchymal stem cells (often harvested from bone marrow) tended to express neuron-like markers on very soft gels, muscle-like markers on intermediate gels, and bone-like markers on stiff gels. The experimental elegance lay in what was held constant: the chemistry of the surface could be the same, but the elasticity changed. When the group inhibited non-muscle myosin II, a key generator of cellular tension, the stiffness effect largely disappeared. The result suggested that the cell was not "smelling" bone; it was mechanically testing its environment and then translating that test into lineage choice.

For a general reader, the jargon can obscure the simple drama: a cell that could become many things commits to one identity partly because of how hard the floor is. This is surprising in the way a good scientific fact should be surprising. We tend to treat fate as something inscribed in DNA and unlocked by chemical keys. Here fate is also shaped by the laws of springs and levers. There is a caution, too: the word "decides" can sound as if the cell is a little mind. What is really happening is a dynamical process. Mechanical input biases signalling networks; signalling networks bias gene expression; gene expression reshapes the cytoskeleton and adhesions; the cell's mechanics change; the loop runs again. Over time, the loop can settle into a stable state that corresponds to a particular cell type.

The clinical appeal is obvious. If mechanics can steer cells, then a biomaterial can be more than a placeholder. It can be a teacher. Imagine repairing a bone defect not by stuffing it with inert filler, but by implanting a scaffold whose stiffness, pore size, and surface texture coax incoming stem cells to lay down mineralized matrix in the right pattern. Imagine healing muscle after trauma by providing an elastic niche that encourages muscle fibres to align and mature rather than scar tissue to dominate. Imagine rebuilding cartilage by offering a material that allows compression and fluid flow in a way that resembles the joint, rather than a stiff implant that rubs and fails.

To appreciate how materials act as teachers, it helps to break down what "mechanical properties" actually mean. Stiffness is the headline measure, but it is not the whole story. A rubber band and a piece of cartilage may feel similarly springy at first touch, yet behave differently over time. Many biological tissues are viscoelastic: part solid, part fluid. Press on them quickly and they resist; hold the pressure and they slowly yield. Cells can sense this time dependence. A stem cell pulling rhythmically on a viscoelastic gel experiences a different feedback than it would on a purely elastic one. The cell is effectively probing not only how hard the material is, but how it dissipates energy.

Then there is topography: the tiny ridges, grooves, and fibres that give a surface texture at the micron and nanoscale. A cell crawling across aligned fibres tends to elongate and orient itself along them, which in turn reorganizes its internal tension. This can influence differentiation, especially in tissues where alignment matters, like nerve, muscle, and tendon. There is also confinement: how much space the cell has to spread. A stem cell forced into a small adhesive island adopts a rounder shape with lower tension; given more area, it spreads, stiffens, and engages different signalling pathways. Even geometry becomes a kind of mechanical cue.

The art, and difficulty, is that these cues do not act in isolation. A hydrogel's stiffness depends not only on how densely it is crosslinked but also on how it swells with water. Changing crosslinking can change how proteins bind. Changing pore size alters diffusion of oxygen and nutrients. Cells, meanwhile, are not passive recipients. They remodel the material by secreting enzymes that cut matrix, by pulling fibres into alignment, by depositing their own ECM. A scaffold is an initial condition, not a final environment. Designing it is closer to planting a garden than assembling a machine: you shape the starting soil and trellises, and living agents do the rest.

One of the most intriguing pieces of the story sits inside the nucleus itself. The nucleus is not a rigid marble; it is a deformable organelle, and its mechanical state can influence gene regulation. A key component of the nuclear "skeleton" is the lamina, a meshwork underlying the nuclear envelope. Lamin-A, one of its major proteins, varies across tissues. Joe Swift and colleagues (2013) reported that lamin-A levels tend to scale with tissue stiffness: higher in stiffer, collagen-rich tissues, lower in softer ones. This makes intuitive sense. A nucleus in a bone cell may need more reinforcement than a nucleus in a neuron. The twist is that stiffness can also *change* lamin-A levels, and lamin-A can influence differentiation. In other words, the nucleus is both a sensor and an effector in mechanobiology, not merely a container for DNA.

The nuclear boundary is not just structural; it is also a gate. Proteins shuttle in and out of the nucleus through nuclear pore complexes, which are selective filters. Evidence has grown that force can tune this transport. When tension is transmitted from adhesions through the cytoskeleton to the nucleus, the nuclear envelope and pores can be mechanically perturbed, altering how readily certain regulators enter. Some cargoes are large and require active transport; others diffuse passively. If mechanical strain changes the balance, it can bias which transcription factors reach DNA and when. The cell's external world, in other words, can influence nuclear access control.

This is one reason biomaterials are now designed not only for bulk stiffness but for how they transmit forces at the scale of single adhesions. Cells grip their environment in discrete points. If a material yields locally under those points, the cell experiences a different micro-mechanical landscape than it would on a material that is globally soft but locally reinforced. Engineers are beginning to tune ligands (the binding sites integrins attach to) and mechanics together, so that the same biochemical adhesion can produce different mechanical feedback. It is a reminder that "mechanical cues" are rarely pure mechanics; they are mechanics delivered through molecular handles.

A concrete example brings the stakes down to earth: repairing a segmental bone defect, the kind that can occur after trauma, infection, or tumour removal. If the gap is large, bone may not bridge it on its own. Traditional solutions include grafting bone from elsewhere in the patient's body (painful and limited), using donor grafts (with risks of poor integration), or using metal implants (strong but not living). Tissue engineering proposes another approach: a scaffold that fills the defect, invites blood vessels, recruits stem cells, and provides an environment that encourages those cells to become osteoblasts, the bone-forming cells.

How would a scaffold "encourage" osteoblasts? One route is stiffness. Bone is stiff, but a scaffold implanted into a body cannot simply be as stiff as bone if it is degradable and porous; it would become brittle or block vascular ingrowth. Designers aim for a compromise: stiff enough to support load and provide mechanical cues, but porous and resorbable enough to be replaced by living tissue. Porosity matters mechanically as well as biologically. A lattice with the right architecture can distribute stresses so that cells experience compressive and tensile strains similar to those in natural bone microenvironments. Some scaffolds are made with gradients: stiffer in regions that bear load, softer in regions that need more vascular invasion, echoing the fact that bone itself is not uniform.

Mechanical stimulation can be layered on top. In vivo, bone cells are exquisitely sensitive to dynamic loading and to the flow of interstitial fluid through the lacunar-canalicular network. That flow produces shear stresses that can activate mechanosensitive channels and drive signalling. In the lab, bioreactors apply cyclic compression or fluid perfusion to cell-seeded scaffolds, promoting more robust bone-like matrix deposition before implantation. The goal is not to "exercise" the tissue in a vague sense but to provide defined mechanical regimes that push mechanotransduction pathways toward osteogenesis.

The historical arc of this idea stretches back to a deceptively simple clinical observation: prolonged immobilization weakens bone. During the early era of spaceflight, the rapid loss of bone density in astronauts made the point unavoidable. Exercise countermeasures were developed partly as a mechanical necessity, not merely a fitness concern. The same logic, translated to tissue engineering, suggests that a scaffold left mechanically idle may not mature into strong tissue, whereas one that experiences appropriate strains may. Bone is a tissue that expects to be used.

Yet mechanobiology also complicates the story with a warning: cells remember. Mechanical exposure can leave lasting imprints on gene expression and chromatin organization, a phenomenon sometimes called mechanomemory. A stem cell expanded for weeks on a very stiff plastic dish may behave differently when later placed into a soft hydrogel intended to mimic brain or fat. Its cytoskeleton, nucleus, and epigenetic state may have adapted to stiffness and not fully reset. This matters because most laboratory cell culture happens on tissue culture plastic, which is orders of magnitude stiffer than most tissues. In other words, the default lab environment can push cells into a mechanically unnatural state before an experiment even begins. Engineers now develop culture systems that maintain cells in more physiologically relevant mechanics, not as a luxury but as a prerequisite for predictable outcomes.

This is where the metaphor of "architects" becomes more than a flourish. Architects do not simply choose a material; they anticipate how loads will travel, how people will move, how the building will age. Tissue engineers, likewise, design not just stiffness but force pathways: how a cell will anchor, how tension will be distributed, how the material will deform over time, how it will degrade, and how cells will remodel it. The chosen design influences whether the body rebuilds functional tissue or defaults to scar.

A striking feature of living architecture is that it is built by many agents with limited information. A stem cell does not know the patient's anatomy. It only knows local cues: the stiffness under one adhesion, the stretch in its membrane, the crowding of its neighbours, the availability of oxygen, the tug of a passing fluid. That local information can still produce large-scale order, provided the cues are consistent and the feedback loops are well tuned. In development, the embryo uses such local rules to sculpt organs. Regenerative medicine tries to recreate a small part of that rulebook in an implantable material.

The future versions of these materials will likely be dynamic. Real tissues do not maintain a constant stiffness; a wound stiffens as collagen is deposited, then softens as it remodels. A developing organ changes its mechanical state as cells proliferate and differentiate. A static scaffold can only approximate such trajectories. Researchers are now making "4D" biomaterials whose properties change over time: gels that stiffen in response to light, materials that soften as enzymes cleave them, networks that release stress gradually, or scaffolds whose fibres realign under load. A time-varying material can, in principle, guide cells through a sequence of states: expansion, migration, differentiation, maturation. It is closer to coaching than to instruction.

There is a final, deeply human motivation behind all this: the wish to heal without trading one problem for another. A metal implant can restore function but may not grow with a child, may not integrate with tissue, may cause stress shielding that weakens adjacent bone, and may eventually need replacement. A chemical growth factor might speed repair but also risk unwanted bone formation or tumour-like growth if misdelivered. A mechanically instructive scaffold offers a different promise: to prompt the body to do what it already knows how to do, but in the right place and with the right architecture.

When a stem cell becomes bone on a stiff matrix, or muscle on an elastic one, it is not obeying a mystical "feel." It is converting force into biology through adhesions, motors, and a nucleus that itself behaves like a responsive material. The scaffold's stiffness and texture become part of the cell's identity-making machinery. In that sense, the most advanced regenerative materials do not merely fill space; they shape decision-making, turning the physics of a material into the biology of a tissue.

## 10.3 The Ethics of Engineered Life

A clever biomaterial can feel like a small miracle: a soft gel that persuades a stem cell to become a neuron, a fibrous scaffold that coaxes alignment like a developing tendon, a microfluidic "organ-on-a-chip" that breathes and pulses with a rhythm set by a pump. The power of such systems comes from the same premise: living tissue is not only chemistry. It is matter under stress, fluid under flow, membranes under tension. Once engineers can write those physical cues into lab-grown tissues, they can do something medicine has rarely managed with such delicacy—shape biology without adding a single drug.

That delicacy, however, is also what makes the ethical questions hard. If tissues can be engineered to respond to force, then their behaviour becomes less like a static sample in a dish and more like a participant in an environment. A mechanically responsive tissue can adapt, compensate, sometimes even surprise its creators. That is wonderful when the goal is to build cartilage that withstands compression or a heart patch that contracts under stretch. It is unsettling when the tissue is neural, or embryo-like, or placed into an animal where it may integrate with circuits that were never meant to include it.

Ethics is often framed as a brake on innovation, but in biotechnology it is more accurate to think of it as a steering system. Without steering, the most sophisticated engine can still crash. The aim is not to declare certain lines of research taboo by instinct, but to ask a few basic questions in a disciplined way: What is being created? What could it become? What harms could occur, and to whom? Who should decide, and using what evidence?

A good place to start is language. “Organoid” sounds like a miniature organ, and in some ways it is. An organoid is a three-dimensional cluster of cells grown from stem cells that self-organizes into structures resembling aspects of an organ: gut, liver, retina, kidney, brain. Many organoids are incomplete models: no full blood supply, no immune system, no complete architecture. Yet even partial organization can matter morally if it enables capacities that are ethically salient, like sensation or pain. Mechanobiology raises the stakes because physical cues often drive maturation. A brain organoid given the right mechanical environment—appropriate stiffness, stress, and perhaps even patterned stimulation—may develop more complex networks than one left in a mechanically impoverished dish.

This leads quickly to a question that sounds like science fiction but is now a serious topic of bioethics: could a brain organoid feel pain? Pain is not just a signal from a nociceptor (a pain-sensing neuron). It is an experience, which in humans involves distributed circuits, learning, attention, and affect. A dish-grown organoid is missing much of what a brain uses to generate experience: a body, sensory organs, hormonal signals, normal development, and the evolutionary scaffolding that shapes a mind. Still, uncertainty is not the same as impossibility. Neural tissue can generate electrical rhythms. Networks can respond to stimuli. If a system becomes complex enough, the moral risk shifts from "almost surely none" to "we should not be casual".

Mechanosensitivity makes this harder to ignore because it provides an engineer's handle on development. Neurons are not indifferent to stiffness. During development, brain tissue is extremely soft, and its mechanical properties change as cells proliferate, migrate, and form connections. Glial cells respond to mechanics too, and they are crucial for shaping synapses and insulating axons. Even the nucleus, with its lamina and its pores, can respond to mechanical strain by changing gene regulation and transport. If an organoid is built in a way that restores these physical contexts, it may become a better model of disease and therapy. It may also, at some level of complexity, become something we should regard with a kind of caution we do not apply to a simple cell line.

The ethical discussion often turns on "moral status": the idea that some entities deserve moral consideration because of what they are or what they can experience. A human adult has high moral status in nearly all ethical frameworks; a bacterium has very little. Animals sit in between, and debates about animal research often hinge on sentience, suffering, and cognitive capacities. Where do organoids belong? Some people's instinct is to sort by origin: human-derived tissue must matter more than animal-derived tissue. Yet origin alone can mislead. A human skin cell is human, but few people think it has the moral status of a mouse. Many ethicists argue that what should matter is not the label "human" but the presence of morally relevant properties: the capacity to feel, to suffer, perhaps to have preferences or a point of view.

Public intuitions appear to track this. When people are asked what makes organoid research troubling, they often focus less on "playing God" in the abstract and more on concrete possibilities: pain, distress, awareness, or the creation of something that cannot be cared for. That instinct is not anti-scientific; it is a request for responsible uncertainty management. If there is a plausible route to suffering, one needs a plan for detecting it, preventing it, and stopping if prevention fails.

Here, a complication appears: how would one detect suffering in an organoid? In animals, behavioural cues and physiology provide clues. In humans, self-report is the gold standard. Organoids provide neither. Researchers can measure electrophysiology, gene expression, network connectivity, and responses to stimuli. But none of these is a direct meter of experience. There is no universally accepted "consciousness biomarker" even in full brains, let alone in reduced models. The ethical response, then, cannot be built on definitive tests. It must be built on graded oversight tied to plausible risk, with conservative boundaries when uncertainty is high.

This is where governance frameworks matter. The International Society for Stem Cell Research (ISSCR) has become a central reference point for international guidelines, and its August 2025 update (Version 1.2) explicitly addresses organoids, embryo models, chimeras, and genome editing, emphasizing oversight and clear experimental endpoints, and prohibiting uterine transfer of human stem cell–based embryo models or attempts to culture them toward anything resembling viability. Such documents do not settle moral questions for everyone, but they provide a shared infrastructure: common categories, review processes, and stopping rules that make research legible and accountable.

A historical anecdote helps explain why such infrastructure is needed. In the United Kingdom in the 1980s, the Warnock Report (1984) responded to the new reality of in vitro fertilization and embryo research by proposing the "14-day rule": embryos should not be cultured beyond 14 days after fertilization. The logic was partly biological (around this time the primitive streak appears, and twinning is no longer possible) and partly political: a compromise boundary that could stabilize a contentious debate and allow valuable research under limits. Whatever one thinks of the number 14, the deeper lesson is durable. When technology shifts what is possible, societies sometimes choose a clear line not because nature provides it, but because governance requires it.

Embryo models and organoids now put pressure on similar lines. "Embryo model" can refer to structures derived from stem cells that resemble aspects of early embryos without being formed by fertilization. They can be scientifically powerful: early development is a period when mechanics and geometry matter profoundly, and embryo-like models allow researchers to study how forces shape patterning. Yet they also raise concerns about moral status and about sliding toward the creation of entities that resemble embryos in ethically charged ways. Here the ethical work is partly definitional: what counts as an embryo, what counts as a model, and which capabilities trigger stronger restrictions.

Chimeras add another layer. If human cells are introduced into an animal embryo or organ, the resulting organism is a mixture. Much of chimera research is directed toward practical goals, such as growing human-compatible organs for transplantation in animals. The moral discomfort often concentrates on the nervous system and the germline. If human neural cells contribute substantially to an animal brain, could the animal acquire altered cognition or suffering? If human cells contribute to eggs or sperm, could the animal produce human gametes? These are not everyday outcomes, but ethics is often about guarding against low-probability, high-consequence events.

Mechanosensitive engineering can amplify the concern because it can promote integration. A graft that is mechanically matched to host tissue is more likely to connect. A neural organoid transplanted into a brain may vascularize, mature, and functionally couple to host circuits more readily if its mechanical environment is tuned to encourage growth and synapse formation. That is precisely what makes such transplants scientifically useful: they can model disease in vivo, test treatments, and explore development. It is also what raises the question of welfare. Standard animal research ethics already requires minimizing suffering, using the fewest animals, and justifying harm by expected benefit. The organoid transplant case adds a further uncertainty: might the graft itself have morally relevant experience, or might it modify the host's experience in unexpected ways?

It is tempting to treat this as a rarefied debate, but it has an immediate practical consequence: what kind of oversight is appropriate? Many jurisdictions regulate animal research under institutional animal care and use committees, while human tissue research and embryo-related work may fall under different boards and laws. Organoid transplantation cuts across these domains. The ethical response is likely to require joint review and more explicit monitoring plans: not only standard animal welfare endpoints, but also criteria tied to neural function, distress, and unexpected phenotypes.

Ethics is not only about what is created; it is also about how therapies reach patients. Mechanomedicine blurs a line that regulators have long relied on: the line between drugs and devices. A pill is regulated as a drug; a stent as a device. But what about a hydrogel that stiffens in situ and releases a protein when stretched? What about an implant seeded with cells that remodel it under load? What about a wearable that delivers patterned vibration to modulate inflammation, coupled to a biomarker-guided algorithm? These hybrids can be called "combination products," but the label does not dissolve the complexity. Each component has different manufacturing standards, testing paradigms, and failure modes.

A drug's central risk is often biochemical: toxicity, off-target binding, metabolic interactions. A device's central risk is often mechanical or electrical: breakage, migration, wear debris, software failure. A mechanoresponsive tissue therapy can fail in both ways at once. If an engineered cartilage implant is too stiff, it may damage surrounding tissue by abnormal load transfer. If it is too soft, it may collapse and inflame. If it degrades too quickly, it may lose function; too slowly, it may block regeneration. If it provokes an immune response, it may fibrose and stiffen, potentially worsening the very mechanical environment it was meant to repair. Regulators and clinicians need test methods that capture this coupled behaviour.

A simple schematic captures the practical difficulty.

*Where mechanotherapeutics get stuck: drug, device, or both?*

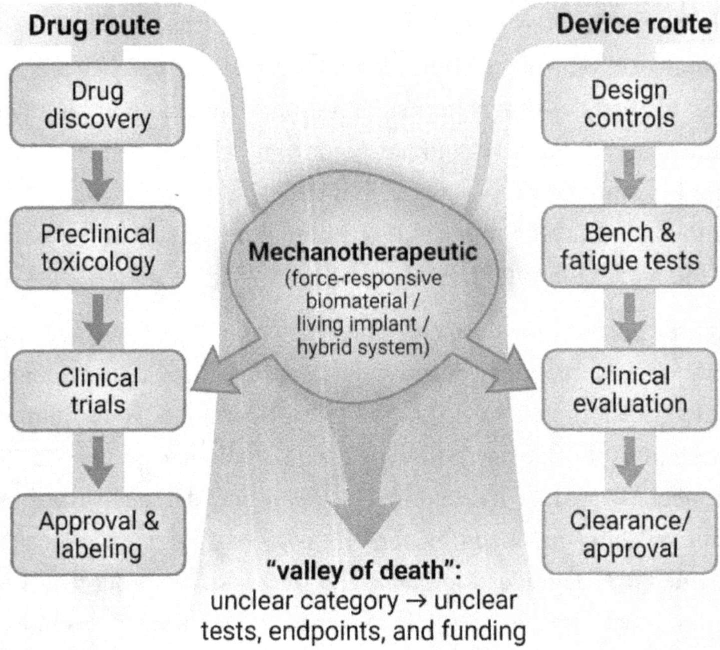


The diagram is deliberately simplified, but the lived reality for innovators is close to its central point: a therapy can be scientifically compelling yet struggle to find the right regulatory "home," and without that home it can be difficult to define the required tests, clinical endpoints, and manufacturing standards. Patients then face a different ethical risk: not premature adoption, but delayed access to something that could help. Ethical governance must therefore balance two kinds of harm: harm from moving too fast, and harm from moving too slowly.

Mechanomedicine also raises questions of justice. Advanced tissue-engineered therapies can be expensive, requiring specialized manufacturing, cold chains, and surgical expertise. If mechanically tuned implants become common, who will receive them? Will access cluster in wealthy hospitals and countries? Will the benefits of reduced disability and improved healing reach those whose work most depends on their bodies: manual labourers, athletes without sponsorship, older adults in under-resourced communities? These are not purely economic questions; they are ethical questions because mechanomedicine is often justified in terms of restoring function and dignity. A therapy that restores walking but only for those who can pay risks widening the gap between medical possibility and lived reality.

There is also the ethics of evidence. When a therapy's active ingredient is mechanics, it may be harder to standardize than a pill. Two clinics can prescribe "the same" rehabilitation program and deliver different strains to tissue because of differences in patient coaching, adherence, and measurement. A mechanoresponsive implant can behave differently in bodies with different activity patterns, inflammation, or scar propensity. This variability makes rigorous trials essential but also challenging. It pushes toward better measurement and transparent reporting: not simply whether a patient improved, but what mechanical regimen was actually delivered, what loads were experienced, and how the implanted material changed over time. Good ethics here looks like good science: clarity about what was done and what was observed.

A practical ethic for engineered living systems therefore has a few recognizable features. It treats moral concern as graded rather than binary: a gut organoid used for drug metabolism is ethically different from a neural organoid with complex activity, and both are different again from embryo-like models. It treats oversight as adaptive: if a model gains new capabilities, review intensifies. It treats patient safety as broader than immediate toxicity: mechanics can harm by misdirecting growth or by destabilizing tissue architecture. And it treats public trust as a fragile resource: secrecy and hype erode it, while clear limits and honest uncertainty strengthen it.

The most sobering thought is also the most hopeful. Mechanobiology teaches that life is extraordinarily good at turning local cues into coherent form. That is why tissues heal at all. Engineering tries to borrow that competence, to provide the cues that let cells rebuild. Ethics, at its best, borrows a similar humility: to recognize that when we create living systems that can respond, adapt, and perhaps one day feel, the responsible stance is neither fear nor triumphalism, but careful attention to what we are bringing into the world and what obligations might come with it.